DATE DUE

Quality, Safety, and Environment

Also available from ASQC Quality Press

Implementing ISO 14001
Marilyn Block

Creativity, Innovation, and Quality
Paul E. Plsek

After the Quality Audit: Closing the Loop on the Audit Process
J. P. Russell and Terry Regel

Understanding and Applying Value-Added Assessment:
Eliminating Business Process Waste
William E. Trischler

Actual Experiences of a CEO: How to Make Continuous
Improvement in Manufacturing Succeed for Your Company
Hank McHale

Mapping Work Processes
Dianne Galloway

To request a complimentary catalog of publications, call
800-248-1946 or visit our web site at http://www.asqc.org

Quality, Safety, and Environment

Synergy in the 21st Century

Pascal Dennis

ASQC Quality Press
Milwaukee, Wisconsin

Quality, Safety, and Environment: Synergy in the 21st Century
Pascal Dennis

Library of Congress Cataloging-in-Publication Data

Dennis, Pascal, 1957–
    Quality, safety, and environment: synergy in the 21st century /
Pascal Dennis.
        p.   cm.
    Includes bibliographical references and index.
    ISBN 0-87389-379-4 (alk. paper)
    1. Industrial safety—Management.   2. Environmental management.
    3. Total quality management.
    T55.D443   1997
    658.4'08—dc21   96-39878
                                                                CIP

10 9 8 7 6 5 4 3 2

ISBN 0-87389-379-4

Acquisitions Editor: Roger Holloway
Project Editor: Jeanne W. Bohn

ASQC Mission: To facilitate continuous improvement and increase customer satisfaction by identifying, communicating, and promoting the use of quality principles, concepts, and technologies; and thereby be recognized throughout the world as the leading authority on, and champion for, quality.

Attention: Schools and Corporations
ASQC Quality Press books, audiotapes, videotapes, and software are available at quantity discounts with bulk purchases for business, educational, or instructional use. For information, please contact ASQC Quality Press at 800-248-1946, or write to ASQC Quality Press, P.O. Box 3005, Milwaukee, WI 53201-3005.

For a free copy of the ASQC Quality Press Publications Catalog, including ASQC membership information, call 800-248-1946.

Printed in the United States of America

 Printed on acid-free paper

American Society for Quality

Quality Press
611 East Wisconsin Avenue
Milwaukee, Wisconsin 53202
Call toll free 800-248-1946
www.asq.org
http://qualitypress.asq.org
http://standardsgroup.asq.org
http://e-standards.asq.org
E-mail: authors@asq.org

To my family—Alexandra,
Eleanor, and Katherine—
who make it all worthwhile.

Contents

Preface

And what is good, Phaedrus, And what is not good—
Need we ask anyone to tell us these things?

—The Dialogues of Plato

I was a second-year chemistry student when I first read *Zen and the Art of Motorcycle Maintenance,* Robert Persig's extraordinary inquiry into values.[1] Persig's quest for "the good" in work and in life touched something deep within me. Writers are society's litmus paper, perceiving changes that are imperceptible to others. Persig felt a malaise in America, a troubling indifference to "the good."

It was 15 years before I became familiar with the work of W. Edwards Deming and Joseph M. Juran, who put a foundation beneath Persig's castle in the air. Deming and Juran had tried to persuade American industry to adopt their "profound system of knowledge"— to no avail. In the heady days after World War II, as America became the world's preeminent industrial power, quality took a backseat to production.

So they took their message elsewhere. The Union of Japanese Scientists and Engineers (JUSE) was eager for new ideas. Deming's lectures to JUSE in 1950 and those of Juran in 1954 ignited what Michael Porter has called the shortest parabola in history.[2] By the 1970s, the United States lost a 50 percent market share in the most important industries of the day.[3] This helped to spark a profound crisis of confidence, a self-examination that is still in progress.

As a chemical engineer specializing in safety and environmental management, I had comparatively little exposure to the quality approach. My colleagues and I were being swept along by a tidal wave of regulation too complex to understand. Its emphasis was on fatalities, injuries, spills, and so on rather than on the systems that produced these outcomes. I used to joke that I never read regulations: It would kill my quality of life.

But we had an underlying sense of unease. Safety and environment laws often seemed to miss the mark or prove counterproductive, and we knew that the connection between compliance and performance was iffy. Over the past 25 years, safety performance in the United States and Canada has remained stagnant or even deteriorated, despite massive investment and regulation.

Performance in environment has been better. Our air, water, and soil are much cleaner than they were a generation ago; belching smokestacks are rare; fish are returning to the Great Lakes. But these improvements have often come despite regulation, not because of it. Environmental laws, now 17 volumes of fine print, have often had perverse effects, well documented by Wildavsky, Bailey, and Howard.[4,5,6] Environmental laws have accomplished much—but not because they were sensible. Spending a trillion dollars in the past 20 years was bound to clean some things up, however inefficiently.[7]

The central theme of this book is that industry requires a new approach to managing safety and environment. The old ways are failing. The paradigms, tools, and techniques of the quality approach, properly interpreted and adapted, are the way out of the wilderness. Companies on the cutting edge are already implementing these ideas, and some are achieving spectacular results. Some plants operate for years without injury to workers or the environment, but these companies are in the minority.

An underlying theme will be the computer revolution and its effect on the quality, safety, and environment professions—indeed, on all professions. As Marshall McLuhan predicted, the computer revolution is having a strongly *abrasive* effect on existing structures in our society. The changes are so pronounced that we numb ourselves in order to cope.

The computer revolution tends to make commodities of the professions. The role of the expert is compromised when access to information becomes universal. Who has failed to notice the relentless pressure on engineers, architects, lawyers, accountants, and other professionals that has been created by the information explosion?

Quality, safety, and environmental professionals must strive to broaden their knowledge and take a multiviewpoint, multidisciplinary approach. They must break down the barriers between their respective specialties and utilize the potential synergy. An accident or environmental incident is really just a "nonconformance."

It is fashionable these days to lump the quality approach with other management fads—as if the work of Deming, Juran, Crosby, Feigenbaum, Taguchi, Ishikawa, Shewhart, and others is the flavor of the month. Management thinking *does* tend to be faddish, and with every publishing season there is a new guru, a new paradigm. But these innovations are generally at the margins, whereas, by contrast, the quality approach is a rich vein of knowledge that will be mined for years.

The stakes in safety and environment are high. Workers' compensation boards across the United States and Canada are tottering on the brink of bankruptcy; public confidence in government's ability to protect our environment is at a nadir; in the professions, there is a sense of futility.

Just before Deming's death in 1993, a colleague of mine, Philip Green, sent him a letter applying his famous 14 points to environmental management. Deming sent back a handwritten note.

He liked the idea.

Intended Audience

This book is written for

- Managers responsible for quality, safety, or environmental performance
- Quality, safety, and environment professionals
- Executives and senior managers with a strategic interest in the future direction of quality, safety, and environmental management
- Lay readers interested in the history of quality, safety, and environmental management and possible future directions
- Students interested in a career in these fields

I have tried to write simply and clearly and to avoid jargon where possible. Parts of chapters 9 and 10, which deal with measurement of safety and environmental performance, assume that the reader has some familiarity with the quality toolbox. The interested reader who is not familiar with these tools is encouraged to acquire a primer. The following are excellent.

- *SPC for Everyone* by John Burr (ASQC Quality Press)
- *The Quality Toolbox* by Nancy R. Tague (ASQC Quality Press)
- *The Memory Jogger II* (Goal/QPC)
- *Waste Chasers* (Conway Quality)

The lay reader may safely skip these sections.

Structure of the Book

Chapter 1 introduces the central theme that quality, safety, and environmental problems share the same root cause and are amenable to the same solutions. It is proposed that worker injuries and environmental incidents are really nonconformances in the quality sense, and, thus, amenable to the quality approach.

Chapter 2 describes a typical company with quality, safety, and environmental problems, and management's efforts to remedy them.

The symptoms common to such companies are identified and a diagnosis is made.

Chapter 3 deals with the quality revolution and includes a short history of managing for quality: the guilds, industrial revolution, era of scientific management, postwar period, and the rise of Japan, as well as the spread of quality knowledge and the contributions of Shewhart, Deming, and Juran; and a detailed description of the quality approach, including its technology, process, and underlying philosophy.*

Chapter 4 describes the current crisis in safety management, including the raw numbers that are threatening to topple workers' compensation boards as well as the inability of regulation to effect improvement. There is also a short history of safety management including the ancients, the guild system, the industrial revolution, the evolution of safety law, and Herbert Heinrich and scientific safety management—and the traditional safety management approach it spawned.

The work of Heinrich and Shewhart is compared. Heinrich's contributions and the weaknesses of traditional safety management are discussed. It is shown that traditional safety management continues to be the predominant approach in industry.

Chapter 5 explores the roots of modern environmentalism. The two archetypal myths that frame much of the current discourse on progress and the environment are discussed. Both the successes and failures of the first wave of environmentalism are described. It is argued that a new approach is needed, one that focuses on the system itself rather than only on results and eventualities—the "end of the pipe."

Chapter 6 discusses the systems approach to management and its application to safety and environment. A generic system structure is proposed and corresponding management activities that more than satisfy the requirements of ISO 9000 and ISO 14000 are described. The "paper wall," a notorious system killer, and leadership, "the wind that fills the sail," are also discussed.

---

*The quality approach is also known as *total quality management* (TQM) and *total quality control.*

Chapter 7 defines a new term, *total safety and environmental management* (TSEM), and discusses its goals, methods, and principles. TSEM entails applying the quality approach to safety and environment. The synergy (and differences) between safety and environmental management are outlined.

Chapter 8 deals with behavior-based safety, a powerful new approach based on behavior psychology. Behavior observation is shown to be the key to upstream measurement—long the Achilles' heel of safety management. Behavior observation's potential application to quality and environment is also discussed.

Chapter 9 deals with the measurement of safety performance and illustrates the application of the quality toolbox to safety. Future directions are suggested.

Chapter 10 deals with the measurement of environmental performance. The absence of standardized approaches, as well as the strategic importance of such measurement to the firm and its various audiences, is discussed. General principles are outlined, as are possible applications of the quality toolbox.

Chapter 11 describes the TSEM system and its interrelated subsystems: the safety management system and the environmental management system. The program menu of each is described in detail. The many common programs are identified, as are programs that have "crossover" potential. Industrial ecology, a powerful new paradigm of industry–environment interactions, and its core tools are described. Full cost accounting, a means of estimating the cost of safety and environment, is also described. Finally, the characteristics of a TSEM company are discussed.

Chapter 12 presents the dimensions of synergy between quality, safety, and environmental management. These are technological, structural, and political.

I hope I have piqued your interest.

Acknowledgments

I would like to thank the editorial staff at ASQC Quality Press for their help and support: Roger Holloway, manager; Susan Westergard, my original acquisitions editor, and Jeanne Bohn, my project editor.

Notes

1. Robert M. Persig, *Zen and the Art of Motorcycle Maintenance* (New York: Bantam Books, 1975).

2. Michael Porter, *The Comparative Advantage of Nations* (New York: Free Press, 1990).

3. *Business Week* (October 1984).

4. Aaron Wildavsky, *But Is It True? A Citizen's Guide to Environmental Health and Safety Issues* (Cambridge: Harvard University Press, 1995).

5. Ronald Bailey, *Eco Scam: The False Prophets of the Environmental Apocalypse* (New York: St. Martin's Press, 1993).

6. Philip K. Howard, *The Death of Common Sense: How Law Is Suffocating America* (New York: Warner Books, 1996).

7. Ibid, 8.

CHAPTER 1

Introduction

Rust never sleeps.

—Neil Young

Synergy Defined

Synergy is made up of two ancient Greek words: *syn,* which means together, and *ergein,* which means to work. Synergy is defined by the *Oxford English Dictionary* as "joint working" and "cooperation." It is a beautiful word, and one that feels right. There is something deeply satisfying about disparate parts working together to achieve a common goal, whether it be in an automobile engine, an excellent organization, or a happy marriage. Much contemporary management discourse asks: How can we make people, groups, organizations, and so on work together?

Kaizen, the key to Japan's competitive success, may be defined as "continual improvement involving everyone, both managers and workers"*—in other words, continual improvement through synergy.

---

*From Masaaki Imai, *Kaizen* (New York: McGraw-Hill, 1986).

Why Synergy Between Quality, Safety, and Environment?

I believe that there is a strong synergy between quality, safety,* and environmental management. Indeed, let me suggest that safety and environment belong under the quality umbrella—no doubt a controversial statement, as turf wars are endemic.

I am not saying that quality professionals have all the answers or that safety and environment professionals are obsolete. Rather, I believe that an occupational injury or an environmental incident is really a nonconformance and is amenable, therefore, to the quality approach.

Let me go further. The computer revolution is rendering the specialist obsolete. When information is readily available to all, what is the value of the narrowly focused expert? The economy of the twenty-first century will place a premium on the broadly skilled, flexible generalist who is able to view a problem from many angles at once and who has many arrows in the quiver. The computer revolution, then, is having an abrasive effect on the professions, pulling some together, rendering others obsolete, but leaving few unchanged. Does it not make sense to anticipate and flow with these forces?

Quality, safety, and environment are three of industry's most pressing concerns. The relentless pursuit of higher quality at lower cost is being driven by ferocious international competition; it is no passing fancy. Joseph M. Juran has predicted that, just as the twentieth century was the century of production, the twenty-first will be the century of quality.[1]

The quality approach, also known as total quality management (TQM) and total quality control, has revolutionized management philosophy and practice. It is that rarity—a paradigm shift—that has allowed resource-poor Pacific rim nations to outcompete the biggest North American companies.

---

*In this book, the term *safety* is used to denote occupational safety and health.

Safety management is being driven by the staggering cost of workplace accidents. Workers' compensation boards across the United States and Canada are in danger of collapse. If normal insurance industry standards are applied, many boards are bankrupt. Thus, safety is a source of bitter division between management and labor unions.

In spite of its potentially devastating effect on share values, environmental management is being driven by its marketing implications and by widespread public dissatisfaction with the state of the planet. Corporate environmental reports are now read by various critical audiences. Environmental performance has become a strategic issue.

The links between safety and environment are obvious. The line separating the plant from its environment is artificial. Consider, in a process that generates a toxic gas by-product, that the gas may be either

1. Confined to the plant

2. Discharged to the environment

3. Captured and neutralized at the source

In the first case, the primary concern is worker safety; in the second, environment; and the third case comprises both. But the root cause is the same in all and is amenable to an integrated approach. Moreover, the safety and environment professions are strongly linked historically: The earliest environmental practitioners were industrial hygienists. Corporations recognize the link and often structure themselves accordingly.

But are there broader links between quality, safety, and environment?

Entropy—The QSE Link

Entropy is the underlying cause of quality, safety, and environment problems. Although some may cavil at this observation, in a problem company the signs of entropy are everywhere. Max De Pree, president of Herman Miller, writes that his most difficult job is the interception of entropy.[2]

Entropy is defined in the second law of thermodynamics, a universal law of nature, as a measure of the disorder or randomness in a system. I will use the term in a looser way. Entropy means chaos, the tendency for things to deteriorate. Entropy has entered folk wisdom as Murphy's Law and pop culture in Neil Young's memorable line quoted at the beginning of this chapter. Quality professionals call it variation in the system.

Thus, left to themselves, things will tend to fall apart. If something can go wrong, it will. Over the years, the second law has been the source of much angst in philosophical circles. Is decay inevitable? Or can entropy be checked, at least for a while?

In fact, the second law tells us that entropy can be slowed down or even reversed by the intelligent application of energy. So what?

Well, a manufacturing plant is a highly ordered system. On a given day, scores of employees must arrive at the right place at the right time, pick up the right tool, perform the right task, and somehow, at the end of the day, produce a product that complies with specifications. The entropic forces working against this system are enormous, and they express themselves most obviously in quality, safety, and environmental problems.

With respect to quality, entropy means increased variation in the process. Process histograms become wider; out-of-control points begin to appear on control charts; and so on. The bottom line is nonconforming product. One of the core goals of statistical process control is the reduction of entropy. Entropy will also flood the workplace with hazards; that is, with unsafe behaviors and conditions. Finally, with respect to environment, entropy will lead to inefficient use of energy and other resources, discharges to the air, water, and soil, and other forms of waste.

Consider an auto manufacturing plant that is beginning to fall apart. Dust and fumes from the welding area drift into the paint shop. The ventilation system designed to control these emissions doesn't work properly. Paint finish is compromised.

The spray painting systems are also malfunctioning. Overspray of paint means higher solvent emissions to the outdoors. Waste water contains high contaminant levels. The welding fumes and paint emissions are also a safety problem: They cause discomfort and reduce visibility, which over time translates into operator error and injury. The paint finish on the cars does not conform with requirements, which means increased scrap or solid waste. Clearly, the plant's quality, safety, and environment problems are closely related.

How does an organization reduce entropy? By applying energy intelligently. Hence, the systems approach.

Management Systems

A management system is an orderly set of components that serves to accomplish one or more goals of the organization. A system's specific goal may be to facilitate the flow of information, to improve quality, to minimize losses due to accidents and injuries, or to reduce environmental impacts. The underlying goal of all management systems, however, is to reduce entropy.

In the absence of a systems approach, companies may be overwhelmed by the exponentially increasing complexity of the business environment. A system channels, organizes, and simplifies complexity. A system provides order, structure, and constancy of purpose. The heart of the famous Toyota production system is standardized work.

The power of the systems approach is now broadly recognized, and there is a powerful trend toward international standards. The ISO 9000 series of standards pertains to quality; the ISO 14000 series pertains to environment; and discussions are underway toward developing a standard for safety management systems (tentatively, ISO 18000).

Let me summarize. Quality, safety, and environment problems have a common root cause—entropy. Entropy, or variation in the system, can be reduced by effective management systems. Safety and environment are really subsets of quality and are thus amenable to the quality approach. There is a powerful trend toward international standards for quality, safety, and environment. Thus, there is a powerful synergy, as yet untapped, except by a small number of elite organizations.

These concepts are summarized in Figure 1.1. The helix represents entropy, the root cause of quality, safety, and environment problems. But it is also the helix of continuous improvement. The triangle represents the management system that controls the chaotic force of entropy. The three sides of the triangle represent the systems triad: quality, safety, and environment.

Notes

1. J. M. Juran, in *A History of Managing for Quality*, ed. J. M. Juran (Milwaukee: ASQC Quality Press, 1995).

2. Max De Pree, *Leadership Is an Art* (New York: Dell Publishing, 1989), 110.

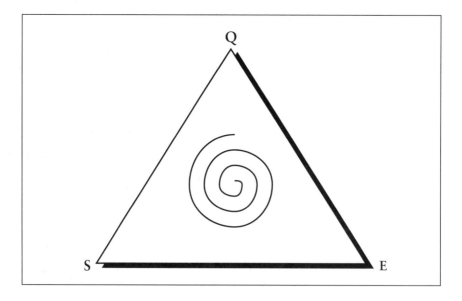

Figure 1.1. Symbol of quality, safety, and environment synergy.

CHAPTER 2

Profile of a Problem Company

Something terrible is happening. People can't do what they're trying to do.

—Charles Bukowski, 1980

Philip Crosby began *Quality Without Tears** with a description of a problem company—and the observation that it does not take Louis Pasteur to make a diagnosis. Here is another company with problems along three dimensions, and this diagnosis also does not require Pasteur. As we shall see, these problems are manifestations of the same root cause, in the same way that steam, water, and ice are different forms of the same material. They just look different.

I have purposely chosen utilities—an industry with which I am familiar and that is undergoing tumultuous change. Little will be left of the old structure in a few years. Thus, it brings into sharp focus the changes that have been sweeping through North American industry. The causes of change are well known—deregulation, international competition, technological change, and so on—and they show no signs of abating.

---

*New York: McGraw-Hill, 1984. This chapter is written with appreciation to Philip Crosby.

Midsize Power and Light* is a producer and distributor of electrical power. A publicly owned monopoly since it was founded at the turn of the century, Midsize's primary business is generating and distributing electricity, as well as maintaining its vast system.

Quality Problems

Midsize charges about 20 percent more than neighboring utilities. The rationale is that the distribution plant is older and thus requires more maintenance. Midsize's rate structure is under attack, however, from major industrial users and from the new governor. In addition, Midsize's monopoly is coming to an end. Soon private generators and neighboring utilities will be allowed to sell electricity directly through Midsize's distribution system. Cogeneration, a competing technology with a substantial cost advantage, is being actively marketed by natural gas suppliers. Many of Midsize's largest customers are requesting proposals.

Price is not the only issue. Midsize has no meaningful database and does not know its customers. There is little dialogue with major customers, let alone residential ones. Residential customers face a difficult computerized telephone system, with long waits and abandoned calls the norm. Reliability has been slipping, too. The distribution network is aging, and power outages are becoming more frequent. Midsize depends on an elite group of troubleshooters to solve these problems.

Safety Problems

Midsize has a poor safety record. Accident frequency and severity rates are several times the industry average. Risk taking is widespread among the workforce, with the result that over the years many workers have been killed. Management has the laissez-faire attitude that

---

*Any similarity between the company described herein and any other company, alive or dead, as they say, is purely fortuitous.

production and distribution of electrical power is simply a dangerous business. Senior management believes that safety problems are caused by worker carelessness.

The most recent fatality involved the installation of a large transformer. A worker was standing on top of it while it was being tested. Because the insulation was substandard, the transformer blew up, killing the worker instantly. Burning transformer fluid contaminated the soil and sewers.

Midsize and three senior officials were charged under both safety and environment regulations. An out-of court settlement for the accidental death was reached. Midsize pleaded guilty to violations and the charges against the officials were dropped. A fine of $500,000 was imposed. Midsize fought the environmental charges and had mixed success: The charges against the officers are suspended but the company was convicted, fined $250,000, and found liable for all related damages.

Reported But Not Incurred

Midsize's workers' compensation premiums are at crisis levels. Midsize suffers from a new phenomenon in workers' compensation insurance: reported-but-not-incurred (RBNI) accidents. Workers regularly file questionable compensation claims and stay off work for extended periods for even the most minor injuries. Most of these are soft-tissue injuries, which are difficult to disprove. Absenteeism rates, incidentally, are twice the industry average.

Environmental Problems

Midsize's environmental performance is also under scrutiny. Several years ago, Midsize made massive investments in pollution control technology that brought emissions into compliance with state regulations.

But the environment has continued to be a thorny issue. Community groups have raised questions about air emissions from the coal-fired plants and their contribution to acid rain and global warming.

Midsize's extensive use of herbicides to control vegetation near power lines has also been criticized. Finally, the public is becoming concerned about electromagnetic fields (EMFs). Recently, a prominent journalist wrote a series of articles linking EMFs to leukemia in children.

Midsize is frustrated by such attention. The company complies with all the applicable laws and has invested in environmental control. The data regarding global warming are inconclusive and may turn out to be a red herring. So why take action now, given all the other problems?

Management Response

Senior managers have all come up through the system and believe in the Midsize way—a deliberate, conservative approach to management. They view the country club atmosphere of the executive offices as their reward. They are rarely seen outside their plush surroundings. But with the thought that perhaps it's time to act more quickly, a quality improvement program is implemented. A policy statement on quality is developed and signed by the CEO, and a rousing kickoff is held.

A dialogue is begun with major customers. Midsize unilaterally reduces its rates by 2 percent. A new telephone system is purchased to reduce phone dead time and abandoned calls, and a customer service hot line is implemented.

Three new directors are hired: one each for quality, safety, and public relations. A number of joint committees are initiated to bring the union into the picture. An absenteeism management program is implemented. An EMF policy statement is developed and publicized. A massive training program is implemented involving all middle managers and supervisors. The purpose is to drive the quality message down to the workers. The union, however, boycotts the program.

Improving safety becomes Midsize's number-one goal. The CEO issues a five-year plan calling for a 10 percent reduction per year in the number of accidents. The plan's key elements are safety training and promotion. All operating employees are given job-specific safety training. Training is reinforced with a promotional program featuring posters, videos, and an electronic bulletin board. Handouts are distributed at safety meetings. The overall theme is "Safety is everyone's responsibility."

Results

In the first year, the results are promising. Questionnaire surveys conducted by the public relations department indicate that there is a heightened awareness of quality and safety. The quality improvement team is progressing well. The accident rate is down almost 15 percent. Even environmental issues appear to be abating, as a result of the poor local economy.

The second year, the wheels fall off. The quality team's recommendations are not effectively implemented. The accident rate begins to drift upward and, by year end, is almost where it started. Worst of all, there is another terrible injury. A worker is blinded while working on energized equipment. Another court case is looming for the company and its senior management.

One of Midsize's most important customers decides to switch to cogeneration, and there is talk that several more are on the way. The state passes a law requiring that Midsize reduce its rates by 3 percent for each of the next two years. Environmental groups once again begin to ask awkward questions.

Symptoms and Diagnosis

Companies with quality problems almost always have safety and environment problems as well. The liability mix will vary with the industry. High-risk industries such as power, forestry, transportation, mining, chemicals, manufacturing, and construction tend to have severe safety and environment liabilities. One mistake can cause a fatality or "total" the plant.

Even industries thought to be comparatively free of safety and environmental risk can and do have expensive problems. These do not entail employee fatalities, but they involve lower back problems and other repetitive strain injuries that are bankrupting compensation boards. These are not spectacular incidents such as Bhopal or Exxon-*Valdez*, but chronic inefficient use of energy, waste of raw materials, and low-level pollution render a site toxic in the long term. It is death by a thousand cuts.

Let's look more closely at the patient's symptoms and see if we can make a diagnosis.

1. **Work conducted deviates from best practice in the industry.** Standard operating procedures, if they exist, are rarely followed. Standards for inputs are also lax. There is no clear understanding of the skills required for different positions, nor is there an effective method of measuring whether employees have the necessary skills. The same applies for other inputs. Thus, employees may or may not have the ability to do their jobs. Raw materials and supplied parts may or may not meet specifications. "Not to worry; we'll make them fit." Everything varies: inputs, the process, and outputs.

Critical behaviors—those that are the common pathway to injuries—are not defined. Critical tasks—those that, if done improperly, are likely to result in a major loss (to people, equipment, materials, environment, or quality)—are not recognized. Nor are critical parts—those parts, components of equipment, instruments, or structures whose failure is likely to result in a major loss. Quality, therefore, becomes a crapshoot; safety and environment, a game of Russian roulette.

2. **The company is focused on the end of the pipe.** Troubleshooting becomes the norm. Ensuring quality requires an extensive field service or dealer network to finish the product, or, as in the case of Midsize Power and Light, continually fight fires.

In safety, end-of-the-pipe thinking involves an emphasis on after-the-fact accident investigations, body counts, and personal protective equipment. It usually coincides with the feeling that workers are careless or incompetent. In terms of environment, such thinking means heavy investment in pollution abatement equipment. End-of-pipe thinking also entails an emphasis on regulatory compliance. In other words, keep the CEO out of jail and the company out of the headlines. Safety and environment (and quality) are believed to be expensive: "We can't afford that level of safety (or quality)."

3. **Management does not provide clear, measurable, trackable goals.** Clear definitions of quality are absent. Customer databases are incomplete or nonexistent. There is only a vague awareness of safety and environment liabilities.

The status quo is the standard. If the process produces 5 percent defective output—then that is the quality standard. If 10 percent of the workforce is injured each year—that is the safety standard. Where environmental legislation exists, the norm is compliance. Otherwise, the standard is the usual amount of solvent sent up the stack or waste-water discharged to the sewers.

Safety and environment performance often are not tracked at all. Morale and labor relations are generally poor. But workers understand: "Never mind what the posters say; they don't care about our safety."

4. **Quality, safety, and environment are not line responsibilities.** Quality is the QC department's job; safety, the safety department's job; and so on. These staff groups rarely have the power to change things. Their function is to shield line groups from hassles from customers, government inspectors, and so on. Line managers are not held accountable for their performance and when things go wrong, they point the finger at in-house quality, safety, and environment professionals. Authority without accountability. Nice work, if you can get it.

5. **Management does not know the cost of quality, safety, and environment.** The cost of nonquality, that is, of doing things wrong then having to fix them, often exceeds 25 percent. Why does management put up with it? Because management is not aware.

Only direct safety costs, such as workers' compensation premiums, are tracked. Indirect costs, which are many times higher, are buried in overhead. Again, management does not know the cost of nonsafety.

Environment costs are tracked in a superficial manner. Pollution control costs sometime appear on income statements. However, balance sheets rarely include environmental liabilities. Environmental cost components—energy, solid and liquid waste, hazardous waste, permits and other compliance costs, public goodwill, and so on—are rarely tracked.

6. **Management denies it is the cause of the problem.** Most senior managers send everybody else to school, set up programs for others to implement, make impressive speeches, and sign policy statements. But the hard work is up to everyone else. Most senior managers believe that quality would improve if only workers paid more attention to their work. Or if middle managers did their jobs.

Safety problems are viewed as the fault of careless workers. The workers' compensation fiasco is caused by poor worker attitudes. Environmental problems are the fault of environmentalists or the media. Problems are addressed as they arise. Often there are random short-term improvements, but these inevitably fade or give way to new problems. Entropy that is chased from one place just turns up somewhere else. Meanwhile, environmental liabilities grow; workers are injured; morale and commitment sink. Before long, profits disappear.

What is the root cause of Midsize's problems? Entropy—too much variation in the system. Why is there so much entropy in this system? Because of the leadership vacuum and the absence of a systems approach.

Are these symptoms familiar? If so, your company probably has serious entropy problems. Let us now discuss the quality revolution— the response to these problems.

CHAPTER 3

The Quality Revolution

To know how to do, know how to be. Each is the expression and the culmination of the other, and an accomplished culture is the harmony of the two.

—Le Compagnonnage, Bernard de Castera[1]

The Fall of the Titans

In 1950, America was an industrial colossus. American products and services dominated world markets to an extent unparalleled since the heyday of the British Empire half a century earlier.

By 1980, America was an industrial basket case, her corporations hemorrhaging market share, her confidence shattered. Figure 3.1 lists industries in which American companies lost 50 percent market share during the 1970s. Many well-known firms went out of business; others, like Chrysler, were saved only by government intervention.

But the other superpower fared even worse. The mighty Soviet Union crumbled almost overnight, leaving an economic, social, and environmental catastrophe in its wake.

The Soviet Union had been a planned society on a mammoth scale. The anticipated benefits included the elimination of internal competition and the "anarchy" of the marketplace. The experiment failed utterly. Though richly endowed with natural resources, the

Automobiles	Food processors
Cameras	Microwave ovens
Stereos	Athletic equipment
Stereo equipment	Computer chips
Medical equipment	Industrial robots
Color televisions	Electron microscopes
Hand tools	Microscopes
Radial tires	Machine tools
Electric motors	Optical equipment

Business Week, October 1984.

Figure 3.1. Industries in which the United States lost 50 percent of market share during the 1970s.

society was unable to provide adequately for its citizens. The failure extended to quality. Excluding military and related equipment, the quality of Soviet products was incredibly poor.

Meanwhile, the Pacific Rim countries, led by Japan—though lacking natural resources and devastated by the war—became major economic powers.

What happened?

The Soviet Union's collapse illustrates the weakness of the nineteenth-century industrial model: out-of-touch leaders trying to run a vast enterprise in a command-and-control style, through a rigid bureaucracy, with little regard to the needs of citizens, employees, or customers. The humbling of so many American corporations illustrates the weakness of the predominant twentieth-century industrial model: the Taylor system.

The Taylor System

Frederick Taylor was a mechanical engineer who had worked as a machinist, foreman, and plant manager in the late nineteenth century. His experiences suggested that the supervisors and workers of his day lacked the education needed to make essential production decisions. At the time, production planning was largely done by master mechanics and shop supervisors. Their approach was empirical, rooted in the craft practices handed down by the guild system. Taylor's solution was to separate planning from production. He assigned engineers and other specialists to plan the work, and left the supervisors and workers the narrow job of executing the plans.

Taylor's system, later dubbed scientific management, achieved spectacular gains in productivity. The resulting publicity stimulated further application of his methods. As a result, the concept of separating planning from execution became firmly rooted in the United States, Canada, and the United Kingdom.

Taylor was a major innovator. His system helped make the United States the world leader in productivity. To some, however, he is an ambiguous figure. His system also resulted in numbing, repetitive work that alienated workers. Indeed, labor unions use the word "Taylorism" with distaste.

Drawbacks of the Taylor System

The Taylor system had a number of drawbacks. Craftsmanship deteriorated as trades were divided into multiple tasks. The feedback loop, central to quality control, was broken. A worker needs at least the following information to produce quality.

- The quality goal
- The actual quality of what is being produced
- A means of adjusting the process in the event of nonconformance with the goal

The rigid division and definition of work under the Taylor system made quality control on the shop floor impossible.

To address these problems, factory managers created central inspection departments whose job was to prevent defective products from reaching customers. Inspection tended to focus on outgoing products. When a defective product did reach a customer, the main question was "Why did you let this get to the customer?" rather than "Why did you make it this way?" Over time, the inspection department came to be known as the quality control department, and a peculiar belief evolved: Quality was the responsibility of the QC department. This idea is still widely accepted.

Joseph M. Juran has summarized the quality management process under the Taylor system in America as follows:

> *Each functional department in the company carried out its assigned function and then handed off the result to the next function in the sequence. This was often called throwing it over the wall.*
>
> *At the end of the sequence the quality department separated the good products from the bad.*
>
> *For defective products that escaped to the customer, redress was to be provided through customer service based on warranties.*[2]

Under the Taylor system, quality took a backseat to production, and for a long time this didn't matter. American corporations could sell all they made. But by the 1980s, it was clear that there were serious problems with the Taylor system.

Effect of the Computer Revolution on the Taylor System

The computer revolution has helped to obsolesce the Taylor system. The digitization of information subverts both centralized authority structures and the command-and-control style. Just as there is no center in a computer circuit, there is no central hub in a computer network or telephone grid system. Any point in the grid is as central as the next. Moreover, the instantaneous transmission of information works abrasively on rigid hierarchical structures. It is also a strong integrating force. When information is widely available and everyone in the organization is connected electronically, the division of labor and separation of planning from production make little sense. But organizations have been slow to adjust to the demands of the new technology. Thus, the Taylor system is an example of the failure of institutions to adapt to the demands of the technology of the day. There have been many such examples throughout history. Successful organizations are those that are able to integrate the new information processing into their decision-making processes.

The quality revolution and the computer revolution are complementary. Both revolutions

- Take a systems approach
- Emphasize integration rather than division of effort
- Favor multiple viewpoints over a fixed viewpoint—hence the team approach
- Are incompatible with a command-and-control management style

Thus, a strong quality culture helps the organization adapt to the demands of the new technology.

Marshall McLuhan predicted that the computer revolution will lead to a renewal in the oral traditions that existed prior to Gutenberg's invention of the printing press. We are already beginning to see in the Internet the fulfillment of his prediction. I am struck by the "orality," so to speak, of Internet bulletin boards. The oral tradition was also the high point of the guilds. Are we seeing the rebirth of the guild approach to quality as well?[3]

Let us take a brief look at the history of managing for quality.

A Short History of Quality
The Guilds

> *I build for men, I build with men, and, what is less*
> *visible as I do this, I build men.*

—Emile le Normand, Locksmith of the Order of Duty[4]

Guilds were monopolistic craft and trade organizations that thrived during the Middle Ages until they were made obsolete by the Industrial Revolution.* The origin of guilds is as old as humanity itself. Corporations of artisans flourished under the pharaohs of the twelfth dynasty, during the golden age of Greece, and under the Roman Empire. When the Roman Empire collapsed, guilds associated themselves with the monasteries. Guilds played a central role in preserving the crafts and trades during the dark ages and in the subsequent rebirth of Western culture.

During the Middle Ages, guilds derived their powers from charters granted by the prevailing authorities, usually the nobility, the church, or the wealthy. The guilds used their monopolies to provide a livelihood and security for their members. They established rules of apprenticeship as well as qualifying standards and tests for advancement to the level of master. The guilds also provided extensive social services to their members, including a basic form of worker's compensation to aid those members who had suffered injury or illness.

Guilds also provided a moral code for their members. Membership in a guild was much more than a way of doing, it was a way of being. Indeed, the root meaning of the word *quality* is way of being (*qualitas*), and this spirit informed the guilds.

---

*For a detailed history, please refer to *A History of Managing for Quality*, edited by J. M. Juran (Milwaukee: ASQC Quality Press, 1995), upon which much of this section is based.

*The true goal of our work is to render service to our
fellow men. This service is of several kinds:*

- *Serving a client (with all that implies concerning
  ethics, relationships of subservience and conflict
  of interests)*
- *Teaching young apprentices;*
- *Loyalty to oneself, and constant personal striving*

*The artisan's work is like a focal point, a magnet to
which men are drawn, and in which a civilization that
honors fine workmanship recognizes its own social
ideal. (Bernard de Castera)*[5]

Quality Management Under the Guilds

Guilds actively managed for quality. They established specifications
governing input materials, manufacturing process, tests and inspections, and finished products. They conducted inspections and audits to
ensure that craftsmen followed quality specifications. In addition, the
guilds applied their mark or seal to add assurance to customers that
product quality met guild standards.

Craftsmanship flourished under the guilds The incomparable
beauty of the gothic cathedrals at Chartres, Notre Dame, Canterbury,
and others is their enduring testament. The skill of the artisan
depended on

- The training received during apprenticeship
- The experience acquired through repetition of the production cycle
- The fact that, in performing the task sequence, the artisan was
  his own customer

The last point deserves elaboration. The best way to learn about
quality is to be your own customer. Suppose a given process comprises
20 tasks. If the same artisan conducts all 20, that person will discover

that task 7 creates a quality problem for task 15. That discovery highlights the problem and ensures that it will be solved. By contrast, when those 20 tasks are performed by 20 different people, the problem is likely to persist.

Guilds and Quality Improvement

Equality among members was a powerful guild policy. To this end, internal competition was discouraged unless it was considered honest competition. Quality *improvement* through product or process innovation was not considered honest competition, and this policy rendered the guilds vulnerable to external competition. The guilds urged local authorities to restrict imports of foreign goods and imposed severe rules to prevent their trade secrets from falling into the hands of foreign competitors. The Venetian glass industry threatened capital punishment to those who betrayed its secrets.

The policy of solidarity stifled quality improvement and was the Achilles' heel of the guild system. The simultaneous development of powered machinery and sources of power in the latter part of the eighteenth century obsolesced the guilds. They continue to exist in Western countries, though their economic importance has lessened. Many guilds maintain their rich mythology and symbolism. The quality revolution seeks to recapture the spirit of *qualitas* epitomized by the guilds.

The Industrial Revolution

The Industrial Revolution and the factory system it spawned obsolesced the guilds and the small independent shops. The main goals of the factory system were to increase productivity and reduce costs. Under the guild system, productivity was low because of the absence of technology, and costs were high because artisans commanded high wages. The factory system was based on mechanization and the division of labor. Adam Smith, in *The Wealth of Nations,* noted the tremendous economies of scale provided thereby.

The guilds could not compete. Eventually, the artisans took their places in the factories.

The Industrial Revolution caused widespread social and economic chaos. Industrial centers grew by orders of magnitude but lacked basic social and sanitary systems. Gangs of craft workers, obsolesced by the factory system, smashed thousands of machines. They were called Luddites, after their apocryphal leader, Ned Ludd.

But mass production at low cost also benefited society by making basic consumer products—food, textiles, clothing—more affordable. This contributed to economic growth and the rise of a large middle class. The growth in affluence and technology created a demand for better housing, furniture and appliances, vehicles, and luxury goods.

Many commentators have noted the similarities between the Industrial Revolution and our time. The computer revolution is also causing widespread social and economic dislocation. Its effects are mitigated in Western countries by strong social safety nets. Fear of technology has spawned many neo-Luddites, particularly in the environmental movement. But the potential benefits of the computer revolution are as enormous and far-reaching as those created by the Industrial Revolution.

Quality Management Under the Factory System

The emphasis on production under the factory system negatively affected quality. The division of trades into component tasks inevitably damaged craftsmanship. The division of work also meant that the feedback loop was broken and workers were no longer their own customers.

Workers were seldom provided with clear quality goals, measuring instruments to assess quality, or the means for improving processes. Mass production meant that each worker produced bits and pieces that were later assembled into finished products. Few saw the process through from beginning to end. Quality suffered, but the productivity gain was so great that the problems could be ignored.

Quality problems could have been addressed in the planning stages of the manufacturing process. As noted earlier, planning was largely done by master mechanics and shop supervisors whose methods were empirical. They had little understanding of the nature of process variation and the resulting variation in products. They could not know how to collect and analyze data to determine whether their designs could actually be carried out; that is, to determine the capability of their processes. Those tools had not been invented yet.

Taylor Again: Planning Is Separated from Production

Which brings us back to where we began: the Taylor system. As we saw, Taylor's solution to these problems was to separate planning from production.

Is Taylor's major premise—lack of worker education—still relevant?

Clearly not. Education levels have risen remarkably. As a result, the major *under*employed asset in the United States, Canada, and the United Kingdom is the knowledge, experience, and creativity of the workforce. Other cultures, notably the Japanese, have found ways of harnessing this resource with outstanding results. Indeed, two of the core goals of the quality revolution are to bring production and planning back together and to harness the worker's knowledge, experience, and creativity.

The Taylor system is still widely used in companies that are sheltered from competition. But the Taylor system has been obsolesced by the quality approach.* I use the word *approach* as a synonym for the Japanese word *do*. Examples of *do*s include: *bushido,* the way of the samurai; *karate-do,* the way of unarmed combat; and *aikido,* the way of mutual harmony. The development of the quality approach featured a steady evolution in thinking, integration of ideas from various countries and industries, and a spread of knowledge across the world. It represents the life's work of the quality *senseis* (Japanese for "masters") and those that have followed them.

---

*To the best of the author's knowledge, this term was first used by David Hutton in *The Change Agent's Handbook* (Milwaukee: ASQC Quality Press, 1995).

The Quality Revolution

The quality revolution began in 1926 at the Bell Telephone Laboratories, where Walter A. Shewhart developed the tools of statistical process control (SPC). These tools comprised

- The new Shewhart control chart
- Use of probability theory to put a scientific basis under sampling inspection
- A plan for evaluating the quality of outgoing products

At the time there was overwhelming demand for Bell automatic telephone switches. Shewhart applied SPC to the manufacturing processes at Western Electric's enormous Hawthorne Works. (Western Electric was the manufacturing arm of the Bell System.) The Hawthorne Works are also famous for Elton Mayo's pioneering work in industrial psychology. Incidentally, two youngsters were employed there then: a statistician named W. Edwards Deming and an engineer named Joseph M. Juran.

In 1931, Shewhart published *Economic Control of Quality of Manufactured Product,** which showed how SPC could be used to improve the performance of systems. SPC enables the observer to distinguish between the following:

- Variation that belongs to the system—deemed a common cause
- Variation arising from a cause outside the system—deemed a special cause

SPC allows the production of enormous numbers of products to demanding specifications *without* mass inspection. But the power of SPC is not limited to manufacturing. It can be used to study and manage any kind of a system.

SPC and systems thinking had hardly any impact on American industry during the 1930s. However, with America's entry into World

---

*New York: Van Nostrand, 1931.

War II, quality control of military products became a major concern, and the statistical techniques of the Bell system were incorporated into government military contracts under MIL-STD-105.

Shewhart's student, Deming, taught SPC to thousands of American engineers and operators. SPC use was widespread among suppliers to the military. In fact, SPC was considered so valuable that it was classified as a war secret. Despite such progress, however, quality still played second fiddle to the top priority—meeting production schedules. The Army-Navy "E" award, coveted by military suppliers, was given not for quality but for meeting delivery schedules.

When the war ended, there was a massive shortage of consumer products. Only U.S. industry, untouched by the war's carnage, could meet this demand. In this booming sellers' market, quality became secondary to sheer volume. Both the use of SPC and management concern about quality declined sharply in the United States.

Deming later ruefully remarked that, by 1947, nothing remained of his work, "not even smoke." Spurred by this disappointment, he began to develop his own theory of management based on quality principles. His theory was still based on systems thinking, but incorporated principles of psychology and human relations. Thus, Deming's new ideas integrated both the hard stuff, such as SPC, and the soft stuff. Deming's 14 points and his seven deadly diseases of American management have become management mantras.

Joseph M. Juran also worked with the U.S. government during World War II. His focus was on managing for quality. His belief was that quality improvement requires the same level of attention that other functions obtain. After the war, Juran became a consultant and author. In 1951 he published his classic, *Quality Control Handbook*,* which introduced the Juran trilogy: quality planning, quality control, and quality improvement. Juran has taught his course "Managing for Quality" to more than 100,000 people in more than 40 countries.

---

*New York: McGraw-Hill, 1951.

The Spread of Quality Knowledge

At the end of World War II, the United States offered assistance to the Japanese in rebuilding their devastated country, including classes in industrial and scientific management. In 1950 the Union of Japanese Scientists and Engineers (JUSE) invited Deming to Japan to teach them his methods. At that time, "Made in Japan" communicated shoddy, unreliable goods. Deming's lectures comprised an introduction to quality control with a focus on statistical methods. Many of Japan's top managers and engineers attended. These sessions made a powerful impression, and they provided the Japanese with a sense of direction and purpose. Deming emphasized that through statistical quality control a resource-poor nation could prosper. An affection developed between Deming and the Japanese people. The lectures were published and sold as a booklet entitled *Deming's Lectures on Statistical Control of Quality*. Deming donated the royalties to JUSE.

Initially, statistical methods were overemphasized and practical results were inconsistent. In 1954 Juran was invited to give a series of lectures on managing for quality. These completed the circle by providing the management tools to go along with the statistical tools. Both Juran and Deming were amazed at the high level and intense interest of the participants. Juran recalls,

> *I had never before encountered so high a degree of participation on the part of upper management. When I gave such lectures in the United States, the audience consisted of engineers and quality control managers. Never before my 1954 trip to Japan, and never since, has the industrial leadership of a major power given me so much attention.*[6]

The Japanese used the ideas of Deming and Juran as the foundation for their system of management, which they continue to refine to this day.

In 1951, JUSE established the Deming Prize, which became Japan's premier award for industrial excellence. This is presented each year in a nationally televised ceremony. In 1960, Deming was presented with the Second Order of the Sacred Treasure by the Japanese prime minister. Juran received the same award in 1981. Today, both Deming and Juran are honored by the Japanese as two of the main architects of their economic miracle.

It has been suggested that Deming and Juran are somehow responsible for the Japanese miracle. This is overly simplistic. They provided American managers with the same tools, but were ignored. Japanese senior and middle managers deserve the bulk of the credit. As Juran has noted, their commitment was such that they would have succeeded no matter what.

Despite being revered in Japan, Deming and Juran remained almost completely unknown in the West until 1981, when CBS television featured Deming in a documentary entitled "If Japan Can, Why Can't We?" Deming continued his work until his death at the age of 93 in December 1993. Juran, now well into his 90s, continues to write, lecture, and consult.

Quality in the Postwar Period

Following World War II, America's dominance as the world's preeminent industrial power was a short-lived reign. We have already seen how the Japanese challenged the adequacy of American quality. There were other powerful challenges as well.

- The growth of consumerism
- The growth of litigation over quality
- Government regulation of quality

The last one also had a marked effect on safety and environmental management.

Government Regulation of Quality

Governments have established laws for quality and safety since the dawn of recorded history. By the 1960s, U.S. government regulation proliferated and widened in scope. A major portion of this regulation was concerned with the health and safety of citizens. New laws were enacted to deal with highway traffic safety, consumer product safety, occupational safety and health, environmental protection, and so on. Each major area required the establishment of new statutes as well as regulatory agencies to administer them.

Unfortunately, much of this legislation failed to address root causes and simply took the most politically expedient course. The National Highway Traffic and Motor Vehicle Safety Act (NHTMSA), passed in 1966, is a useful illustration of the process. Overwhelming evidence showed that the limiting factor in traffic safety was the motorist. In particular, it was clear that

- Alcohol was an important factor in fatalities.

- Young drivers accounted for a disproportionate number of accidents.

- Most accidents involved speeding, tailgating, or other forms of risky behavior.

- Most motorists did not buy or wear safety belts, even when mandated by law.

Vehicle failure, by contrast, was rarely a contributing factor. Nonetheless, despite the data, the NHTMSA ignored the main problem—improving the performance of the motorist. Instead it concentrated on setting numerous standards for vehicle design. However, the gains in safety were marginal, and the added costs, which the consumer ultimately bore, ran into the billions.[7] This pattern has repeated itself many times over the past three decades.

A second important concern has been the inability of regulatory agencies to grasp the Pareto principle that most of our problems can be traced to a few causes. The number and variety of complaints is enormous and an agency that attempts to deal with them all will become bogged down. Unfortunately, most regulatory agencies continue to ignore the Pareto principle of the vital few. As a result, the costs and benefits of regulation continue to be sources of often bitter division between industrial producers and regulators. During the 1960s and 1970s, the political climate in the United States generally favored the regulators. During the past 15 years or so, the climate changed and the trend is now toward justifying the cost of regulation. As a result, many agencies face severe budget cuts.

"The Quality Dikes"

During the twentieth century, public suspicion and fear of technology have grown to epidemic proportions. Indeed, "doom haunts the end of the twentieth century," as Ronald Bailey has noted.[8] This is human nature, in part. Our oldest stories and myths express an atavistic sense of guilt surrounding the acquisition of knowledge. Adam and Eve are expelled from the Garden of Eden for eating from the Tree of Knowledge. Prometheus is condemned to eternal suffering for stealing fire from the gods.

Public fear of technology is also rooted in the many disastrous ends to which technology has been applied during this century. These include two catastrophic world wars, genocide, the detonation of the first atomic bomb, and the arms race. In fact, as we shall see, the apocalyptic turn the environmental movement has taken is informed by such psychology.

On the other hand, technology confers magnificent benefits upon society. The twentieth century has also been the age of miracles. Hunger, disease, and poverty have been eradicated from large parts of the globe. Even in the bleakest areas, life is far better than it was a century ago and will continue to improve. But technology also makes society dependent on the quality of the goods and services it provides. Hence Juran's concept of "life behind the quality dikes." Like the

Dutch, who have reclaimed their land from the sea, we secure benefits from technology. And, like the Dutch, we require good-quality dikes to protect us from potential disasters. The resulting legislation is often criticized by manufacturers, but consumers have shown that they are willing to pay for the equivalent of good dikes.[9] Let us now look more closely at the quality approach.

What Is the Quality Approach?

The quality approach may be thought of as a diamond with many facets. Turn it in the light and different colors shine through. The different facets may be summarized under the quality triad.

1. Leadership

2. Measurement

3. Participation

Leadership and participation are the soft stuff; measurement, the hard stuff. You need both to succeed. Measurement without leadership means that the statistics won't be integrated into decision making. Having leadership and participation without measurement is akin to navigating without a compass. The captain and the crew may be excellent, but they will grope in the dark without data.

The core process in the quality approach is continual improvement through statistical tracking of objectively defined upstream measures of performance. Everyone in the company must participate in improvement efforts.

Defining Quality

The quality senseis have defined quality as follows:

- "Quality is fitness for use." (Juran)

- "Quality is conformance to requirements." (Crosby)

- "A product or service possesses quality if it helps somebody and enjoys a good and sustainable market." (Deming)

Perspective is the fundamental aspect in any definition of quality. The quality of a specific product or service may be reviewed from a customer's perspective or from the supplier's. Both views are useful, but the customer's view is paramount. Management must start to measure quality as seen by the customer, not solely as seen by the organization.

Quality as a Way of Being

Quality is also about values. Its roots are *qualitas,* the Latin word meaning "way of being," and *arete,* the ancient Greek word meaning "excellence." To the ancient Greeks and Romans these words implied a sense of duty—duty not to others, but to oneself: the duty to strive toward excellence, the duty to be the best that one can be. The concept of continual improvement, central to quality management, captures this philosophy perfectly.

In *The Greeks,* H. D. F. Kitto had this to say regarding *arete*/quality:

> Arete *implies a respect for the wholeness or oneness of life, and a consequent dislike of specialization. It implies . . . a much higher idea of efficiency, an efficiency which exists not in one department of life but in life itself.*[10]

Quality, defined thus, is a universal human value, something we instinctively understand and to which we are drawn. It has found expression in great civilizations throughout history. In ancient Egypt, it was known as *ma'at;* in India, *dharma,* in China, the *tao.* Deming's impact is as much a product of the morality inherent in his system as of the elegance of his statistical tools.

The Goals and Principles of the Quality Approach

The quality approach represents a revolution in management philosophy and practice. It has obsolesced both the nineteenth-century model and the Taylor system. But, unlike other revolutions, this one is largely invisible. Differences in management practice are harder to spot than differences in equipment and technology. Here are some of the goals of the quality approach.

- To eliminate waste by reducing variation in the key processes
- To put the company in touch with the "voice of the customer" so as to allow the design and provision of products that benefit society
- To break down barriers within an organization so as to allow everyone to work on continual improvement

The ultimate goal of a quality company is to benefit its employees, its customers, and society as a whole.

What are the underlying principles of the quality approach? The quality approach

- Is led by senior management
- Puts everyone to work
- Is system-oriented
- Focuses on upstream prevention
- Aims at kaizen or continual improvement
- Focuses on long-term goals
- Bases management decisions on data
- Requires an integrated approach*

It Is Led by Senior Management

Pursuing quality requires a management transformation. This can happen only if senior management is committed. Managers must lead through personal example. They must roll up their sleeves.

---

*I am indebted to David Hutton, who articulated many of these points in *The Change Agent's Handbook* (Milwaukee: ASQC Quality Press, 1995).

It Puts Everyone to Work

Every person in the organization must have the opportunity to develop and to contribute to the transformation. In particular, the experts—frontline workers—must be involved in problem solving, planning, and decision making. This self-directed, self-motivated way of working requires extensive training and education.

It Focuses on the System

The organization is a system made up of many processes, and a company improves by understanding and changing its key processes. According to Deming, management controls 85 percent of the system and the individual employee only 15 percent. Focusing only on individuals, therefore, is fruitless. Most of the error and waste are inherent in the system and beyond the control of employees. Therefore, management must guide efforts to improve the system, which requires a good understanding of the relationships between the process and the results achieved through study, research, and experiment. The quality toolbox provides the necessary techniques.

It Focuses on Upstream Prevention

Eliminating error and waste requires upstream prevention rather than end-of-the-pipe reaction. The root causes of problems must be eliminated. Detecting and rectifying problems after the fact is insufficient. For example, measuring and treating high levels of chromium in waste water does little for the environment or for the company. High chromium levels mean that there is a system problem upstream, which is where we must look for the solution. If the system is not changed, the results will not change.

It Aims at Kaizen

There is always a way of providing a better product or service at the same or lower cost. For example, it is possible for a company to reduce the risks to which its employees are exposed, or to reduce the impact of its activities on the environment. According to Philip Crosby, the performance standard is zero defects. Thus, the goals in safety and environment should be zero accidents and zero discharges.

Why continually improve? Why not take it easy, especially if one is way ahead of the competition? Because the goalposts are constantly shifting; customers' needs are always changing; technology is constantly advancing, thus making better ways possible; and the competition is always improving. The kaizen principle exemplifies quality as a way of being.

It Focuses on Long-Term Goals

One of Deming's seven deadly diseases is short-term thinking and its emphasis on short-term profits. To stay in business, companies must plan for the long term. It is relatively easy to maximize short-term results by cannibalizing assets, ignoring maintenance, and so on, but it is a recipe for disaster. Destructive, job-hopping executives who use this strategy leave a trail of devastation behind them.[11]

Management Decisions Are Based on Data

Decisions are made based on data and an understanding of the cause-and-effect mechanisms at work in the system. This does not mean ignoring soft issues, discounting the judgment of experienced people, or pretending that unquantified factors do not exist.

For example, accurately estimating the cost of safety and environment liabilities may be impossible. But these losses can be crippling and should not be ignored because of a lack of precise data. Full cost accounting can be used to estimate such less tangible costs.

Management by fact does mean using data in a methodical way to understand and solve problems. It means obtaining data on important issues like employee morale and customer satisfaction. It means facing reality, even if it is unpleasant.

The last characteristic is also the most important.

It Requires an Integrated Approach

A collection of programs is not a system. A management system must include a coherent set of values, policies, and mechanisms that are

- Accepted and applied throughout the organization
- Integrated into the normal operation of the business
- Applied consistently

An effective management system will also take into account both the formal structures of the business and its culture.

The Quality Approach Is Not a Panacea

The quality approach can benefit every process and maximize the performance of a system, but it is not sufficient by itself. The quality approach cannot save a company if its competitors develop superior processes or its products are obsolesced. It cannot prevent corporate suicide through lack of market insight or inadequate investment. Even with the quality way, you cannot operate an accounting firm without skilled and experienced accountants. You cannot run a great auto repair shop without talented mechanics, and you still need to spend millions regularly to keep your paint line up-to-date.

The Quality Toolbox

The quality toolbox comprises tools and techniques developed over the past 70 years. These tools help to

- Identify and quantify problems
- Harness the problem-solving skills of everyone in the company
- Evaluate, control, and improve key processes

The core tools are the plan-do-check-act (PDCA) cycle, the original seven quality tools, and the seven planning tools. New tools continue to be developed.

A caveat before we begin. Unless the organization is passionately committed to the quality approach, the quality tools will fail. The gardening metaphor is apt: Management must provide fertile soil, water, and sunlight in order for the quality tools to take root.

The Plan-Do-Check-Act Cycle

Current status, desired status, countermeasures.

—Management mantra

The PDCA cycle is usually credited to Deming, even though he repeatedly cited Shewhart as its originator.[12] It is deceptively simple.

- *Plan* what you want to accomplish over a given period of time and what you need to do to get there.
- *Do* what you plan on doing. Start on a small scale with a pilot program.
- *Check* the results of what you did to see if you achieved your objective.
- *Act* on the information. If you succeeded, standardize the plan; if not, continue with the cycle to improve.

The essence of the PDCA cycle is

- An objective assessment of the current status
- A clear understanding of the desired state and the gap between the two
- A plan to bridge that gap

PDCA is easy to understand. Would that it were easy to do. It is based on a profound problem consciousness that requires extensive investigation. Toyota's "go see" principle is a laconic expression of this idea. Before formulating a plan you must "go see" in order to grasp the situation.

PDCA often faces organizational barriers. Honestly assessing the current status can be a risky career move. In many organizations, it is bad manners to go to the root of things. The check phase is also misunderstood. Checking requires clearly defined measures, checkpoints, and a simple progress rating system. Checking requires that you frequently "go see." PDCA is not for the faint-hearted.

The Seven Quality Tools

There is a saying that "hard data drive out soft data." The absence of data is a universal management problem. Another is poor understanding of the science of variation. The original seven quality tools are primarily data management tools. They comprise

- Flowcharts
- Check sheets
- Ishikawa diagrams (also known as fishbone diagrams)
- Pareto diagrams
- Histograms
- Run charts
- Control charts

These tools are the heart of SPC and allow you to improve your systems. They can be broadly applied, as in an environmental system, or more narrowly utilized, as in a chemical process. You may define a system as you see fit. SPC is beginning to be used effectively in the service industries. Let us briefly look at some of these tools.

Process flow diagrams. You need to know how a process works before you can control it. What are the inputs and outputs? What operations are conducted? Where are the delays? Once you have defined the system, you can start to improve it by identifying unnecessary steps, bottlenecks, and so on.

Check sheets. Check sheets are an information-gathering tool. Once the key measures of a process have been determined, devise a simple form to capture the data and train operators in its use. Collect the data regularly and analyze them using the quality toolbox.

Run chart. A run chart turns data into information by plotting a variable over time. It makes it easier to visualize what is happening in the process and to detect trends.

Ishikawa diagrams. An Ishikawa diagram is a method of organizing information about a problem or a goal. It works best in the context of a team brainstorming session. Start with a clear definition of a problem or goal. Does everyone understand it? Put the definition in a box on the right side of a page.

Now draw a line to the left and put diagonal arrows on it as shown in Figure 3.2. Label the arrows using PEME—people, equipment, materials, environment. Sometimes you will want to add other items, such as a system or measurement. Now you or the team can begin to fill in the things that you think will cause the problem or help to achieve the goal. There are only a few rules.

- Write down everything people suggest. Do not judge.

- The information and not the form is important.

- Everyone on the team must contribute.

Once you have come up with a list of key factors, you can hold a vote and determine which are the most important. Use Pareto analysis to do this.

Pareto charts. Pareto charts are based on the Pareto principle. Pareto was an Italian economist who, in the late 1800s, found that most of the wealth in Italy was owned by a few of the people—which, for some reason, was news. Today we find that the principle applies universally: Most of our problems can be traced to a few causes. This is also known as the principle of the critical few. Many efforts fail because

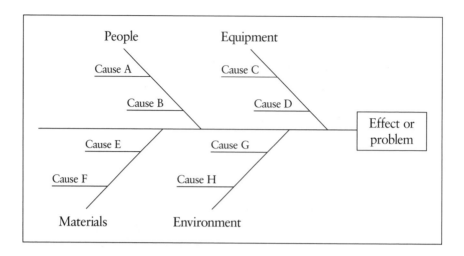

Figure 3.2. Ishikawa diagram.

too much is attempted. Energy is dispersed every which way and nothing substantial gets done, whereas it is far better to focus scarce energy and resources on the critical few.

The 80/20 rule states that 80 percent of the problems come from 20 percent of the causes. The usual way to show this is with a bar graph, in which the height of each bar shows the number of times that the cause occurs. The causes are then placed in order from highest to lowest. Thus, Pareto charts prioritize problems.

Histograms. A histogram is a picture of the variation in a process or a product. Histograms tell us what the distribution of a sample looks like, what the central tendency is, and the spread of the data. Most histograms are bell-shaped, representing the famous normal distribution, so-called because that's what natural distributions normally look like. There are a number of ways of measuring the central tendency: the mode (or highest bar), the mean or average, and the median. Usually, we use the mean. The spread in the data is measured by the range and, most commonly, by the standard deviation.

Control charts. A control chart is a moving picture of variation in a process—"the process talking to us."* Control charts are used to control, stabilize, and improve processes. The control chart takes parameters from the histogram (mean and range) and plots them over time. Statistical formulas are used to plot upper and lower control limits. Variation within control limits is said to be due to common causes, that is, to causes within the system. Variation outside the control limits are said to be due to special causes, that is, causes outside the system.

For example, consider a control chart of the accident frequency rate at a plant over a five-year period. The common causes in this system are the equipment, workers, procedures, managers, plant environment, and so forth. Special causes might be a new piece of equipment, new workers or managers, tense relations due to an impending strike, a new training program, or the like.

The trick is to identify and control both common and special causes such that the accident rate continually drifts down. Control charts carry useful information about changes, both good and bad, in the process. A key concept: You should measure upstream variables because

- Upstream variables, by definition, provide an early warning of breakdowns.

- The earlier in the process you identify an error, the easier and less costly it is to fix.

Scatter diagrams. Scatter diagrams are often used with the seven quality tools. Scatter diagrams show how process variables are related. They are used to identify and evaluate possible cause-and-effect relationships. For example, there might be a relationship between the age of workers and a given type of accident, or between the time of day and a type of incident. Scatter diagrams allow you to investigate these relationships.

---

*Attributed to I. W. Burr.

The Seven Planning Tools

Formal research on the seven planning tools began in 1972, as part of the meetings of the Japanese Society of QC Technique Development. It took several years of research to refine the new tools. The original seven quality tools were adequate for data collection and analysis, but the new tools allow for more identification, planning, and coordination in problem solution. They are listed as follows:

- Affinity diagram
- Tree diagram
- Process decision program chart
- Matrix diagram
- Interrelationship digraph
- Prioritization matrices
- Activity network diagram

We will not review these tools here. Shigeru's *Management for Quality Improvement—The Seven New QC Tools,*[13] Brassard's *The Memory Jogger Plus+,*[14] and Tague's *The Quality Toolbox*[15] are excellent resources.

Problem Solving

The quality tools were invented to solve problems. The following six-step cycle illustrates the problem-solving process and the appropriate tools.

1. **Identify problems; select one to work on.**
   brainstorming, Pareto diagrams, the seven planning tools

2. **Define the problem. If it is large, break it down and solve each component one at a time.**
   Ishikawa diagrams, flowcharts, check sheets, Pareto diagrams

3. **Investigate the problem. Collect data and facts.**
   check sheets, histograms, control charts, scatter diagrams

4. **Analyze the problem. Find all the possible causes and prioritize.**
   brainstorming, check sheets, Ishikawa diagrams, histograms

5. **Solve the problem. Choose from available solutions and implement.**
   brainstorming, check sheets, Pareto diagrams, the seven planning tools

6. **Confirm the results. Collect more data and keep records to ensure that the problem is fixed.**
   check sheets, control charts, Pareto diagrams, histograms

Other Useful Concepts

Let's take a brief look at each of the following:

- The team approach
- Benchmarking
- ISO standards

The team approach. Strong teams facilitate the quality approach. They provide many advantages.

- Teams are collectively smarter and have more experience than individuals.
- Teams can examine a problem from many viewpoints at once.
- Multifunctional teams can circumvent organizational barriers.

It should be noted that team decision making takes time, and teams comprised of like-minded individuals are prone to the groupthink syndrome. Establishing effective meeting habits is important to team success. Strong interpersonal and communication skills in a group greatly enhance improvement efforts.

Benchmarking. Benchmarking is the search for the best practices in a business activity. It involves systematically studying how others tackle some specific process and recording the levels of performance they achieve. This information provides valuable insights and allows more challenging goals to be set.

Benchmarking can focus on both process and results. Results benchmarking alone is of limited value; it provides neither insight into why performance differs, nor ideas for improvement. For example, results benchmarking the company with the best environmental record in an industry without also benchmarking its processes is likely to be a fruitless exercise.

ISO standards. ISO 9000 is a series of international standards that define the elements of a quality management system. It focuses on the procedures and documentation required to demonstrate conformance to specifications for the purposes of a contract. ISO 9000 standards are widely used as a basis for formal registration. Organizations can undergo periodic audits by an accredited registrar, and thus become listed in a published register. Applying these standards can help improve documentation and establish process discipline.

There is a significant limitation: The ISO 9000 series calls for evidence of a system, but not for evidence that the system is succeeding. Thus, purchasing from an ISO 9000–registered supplier does not ensure that the product or service will be satisfactory.

ISO 14000, a series of standards dealing with environmental management systems, has also been developed. It is not yet widely used but is expected to gather momentum. ISO 14000, unlike ISO 9000, explicitly requires evidence of continual improvement. Work is also under way toward a standard for safety management systems.

Summary

The core process of the quality approach is continual improvement through statistical tracking of objectively defined upstream measures of performance. Everyone in the company must participate in improvement efforts.

Under traditional safety and environmental management, by contrast,

- Workers are excluded from improvement efforts, and often are blamed for accidents and environmental incidents.
- There are no meaningful upstream performance measures.
- Statistical literacy is rare.

Traditional safety management had strong success for half a century. But for the past 30 years, the ship has been adrift without a compass. Traditional environmental management has also had strong success over the past 25 years. But, compelled by onerous legislation, its goals have been fixated on the end of the pipeline, even though the real gains are upstream.

It doesn't have to be that way.

Notes

1. Michael Deleforge, in *A History of Managing for Quality*, ed. J. M. Juran (Milwaukee: ASQC Quality Press, 1995), 411.

2. J. M. Juran, in *A History of Managing for Quality*, ed. J. M. Juran (Milwaukee: ASQC Quality Press, 1995).

3. Marshal McLuhan, *Essential McLuhan*, ed. Eric McLuhan and Frank Zingrove (Concord, Ont.: House of Anansi Press, 1995).

4. Michel Deleforge, in *A History of Managing for Quality*, ed. J. M. Juran (Milwaukee: ASQC Quality Press, 1995), 411.

5. Ibid.

6. J. M. Juran, in *A History of Managing for Quality*, ed. J. M. Juran (Milwaukee: ASQC Quality Press, 1995), 539.

7. Ibid., 571.

8. Ronald Bailey, *Eco Sam. The False Prophets of Ecological Collapse* (New York: St. Martin's Press, 1993).

9. J. M. Juran, in *A History of Managing for Quality*, ed. J. M. Juran (Milwaukee: ASQC Quality Press, 1995), 579.

10. Robert M. Persig, *Zen and the Art of Motorcycle Maintenance* (New York: Bantam, 1974), 340.

11. Charles M. Kelly, "If Disaster Strikes, There's Shelter at the Top," *The Globe and Mail,* 10 October 1988.

12. W. Edwards Deming, *Out of the Crisis* (Cambridge: MIT Center for Advanced Engineering Study, 1986).

13. Nizuno Shigeru, *Management for Quality Improvement—The Seven New QC Tools* (Cambridge: Productivity Press, 1988).

14. Michael Brassard, *The Memory Jogger Plus+* (Methuen, Mass.: Goal/QPC, 1994).

15. Nancy R. Tague, *The Quality Toolbox* (Milwaukee: ASQC Quality Press, 1995).

CHAPTER 4

The Crisis in Safety Management

The first duty of business is to survive, and the guiding principle of business economics is not the maximization of profit—it is the avoidance of loss.

—Peter Drucker

Why Safety Management Is Important

More Americans have been killed at work than have been killed at war—by a long shot. At least 1,400,000 Americans have been killed in workplace accidents in the twentieth century alone.* This figure does not include deaths by occupational disease, almost certainly a very large number. By contrast, since the American War of Independence, about 1,300,000 Americans have been killed in combat.[1] An analysis of workplace injuries compared to wartime injuries leads to a similar conclusion.

The social and economic costs of this carnage have never been fully tabulated. Its corrosive effect on worker morale and labor–management relations cannot be underestimated. How can there be workplace harmony with such a legacy? An amputation, an occupational disease, a fatality are never forgotten.

---

*This is based on the National Safety Council's data from 1933 to 1994. Data prior to 1933 are unreliable, so it is assumed that the number of deaths for each year from 1900 to 1933 was 15,000, the same as in 1933. Deaths by occupational disease are unavailable and are not counted.

But for many North American managers, worker safety is a non-issue. I once heard a manager boast that "real managers don't do safety meetings." He was a fool, but this opinion is still widespread. North American managers misunderstand the visceral effect of such statements on the workforce, and cloaking this attitude in high-sounding policy statements and other platitudes is no better. Workers know.

At elite companies, by contrast, safety is a core value, on par with quality and production. Safety is an integral part of the world-famous Toyota production system, for example. Its cooperative, participative approach would be impossible if management were indifferent to the well-being of team members. Indeed, according to Eiji Toyoda, "Safe work is the door to all work. Let us pass through this door every day." This metaphor speaks volumes.

Regulation Explosion—Performance Implosion

Great progress has been made in reducing the number of industrial fatalities. From 1933 to 1994, worker fatality rates decreased by a factor of 10 (Figure 4.1). Unfortunately, the news is not so good when it comes to disabling injuries. Since 1970, injury frequency and severity rates have actually increased after decades of steady improvement. This despite massive investment, attention, and legislation (Figure 4.2).

An ironic result: The explosion in health and safety regulation leads to an implosion in performance. Such results are fueling a powerful trend toward deregulation. In the United States, the very existence of the Occupational Safety and Health Administration is being questioned.

The Workers' Compensation Crisis

Safety management in the United States and Canada is in a state of crisis. If normal insurance industry standards are applied, many workers' compensation boards are technically bankrupt. Mushrooming medical costs are one of Deming's seven deadly diseases of western industry.

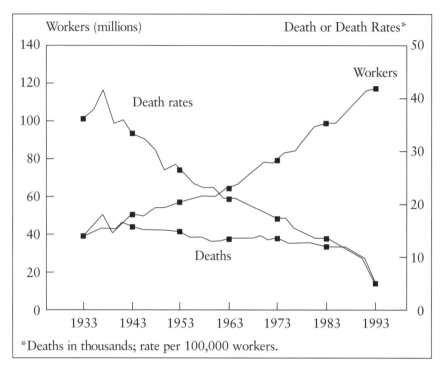

*Deaths in thousands; rate per 100,000 workers.

The National Safety Council (1995), *Accident Facts,* 1995 ed. (Itasca, Ill.: National Safety Council). Used with permission.

Figure 4.1. Workers, deaths, and death rates, United States, 1933–1994.

Workers' compensation costs are a source of bitter division between industry and labor unions. Industry charges systemic fraud and abuse. Insurance companies have traditionally built an incurred-but-not-reported (IBNR) factor into loss-reserving practices. IBNR funds losses that have occurred but have yet to be "claimed" because of a delay in either reporting or manifestation of the injury. Today, a new phenomenon exists: reported but not incurred (RBNI).[2] RBNI is largely related to soft-tissue injuries such as back and repetitive strain injuries (RSI). Claims—the dollar value of accidents—are largely discretionary, a function of employee perception and attitude.

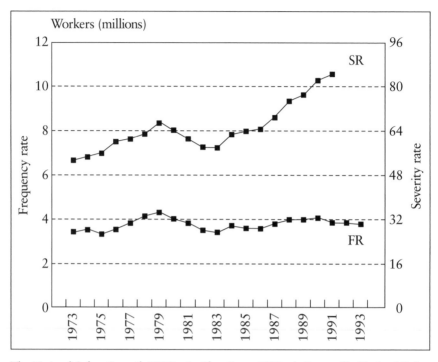

The National Safety Council (1995), *Accident Facts,* 1995 ed. (Itasca, Ill.: National Safety Council). Used with permission.

Figure 4.2. Frequency and severity rates in the United States, 1973–1994.

Labor unions meanwhile believe that the RBNI issue is grossly exaggerated and yet another example of management's blaming the victim and avoiding responsibility for safety. They decry the injustice involving injury to good workers trying to make a living, whose incomes are cut to a fraction of what they were, but who are denied the right to sue.

The raw numbers are distressing. In 1994, in the United States, there were 3,500,000 disabling injuries, each costing an average of $29,000. Because of injury, 125,000,000 work days were lost in 1994. Future time lost because of 1994 injuries is estimated to be about 60,000,000 days. The total cost of work injuries and illnesses in 1994 was about $120.7 billion.

We are not used to dealing with such gargantuan numbers. Let me put them in context.

- About 14,000 workers suffer disabling injuries every day in the United States.

- Assuming an after-tax profit margin of 10 percent, American industry has to generate gross revenues of $1.2 trillion to pay the cost of workplace injuries and illnesses.

Yet there are organizations whose safety performance is hundreds of times better than the industrial average. There are large high-risk plants that go for years without an injury or environmental incident.

Before we attempt a diagnosis, let us take a brief historical tour.

A Short History of Safety Management

> *Each master . . . has in great measure the health and happiness of his work people in his power . . . let benevolence be directed to the prevention rather than to the relief of the evils.*
>
> —Charles Thackrah
> Eighteenth-century British physician

Early History

Safety is not a new management responsibility. Safety laws have existed since the dawn of recorded history. The code of laws of the Babylonian king Hammurabi (circa 2200 B.C.) prescribes punishment of overseers for injuries suffered by workers. Safety laws are also prescribed in *The Pentateuch,* the first five books of the Old Testament, traditionally ascribed to Moses.

The great physicians of Greece and Rome were aware of the diseases of the occupations. In the fourth century B.C., Hippocrates recorded the symptoms of lead poisoning among miners. Five centuries later, Pliny the Elder referred to the dangers inherent in dealing with lead and asbestos and advised laborers to wear protective masks when working in high-dust areas. In the second century A.D., Galen recognized the dangers of acid mist to copper miners.

In general, though, the ancients showed little concern for worker safety and health. Human life was cheap; life spans were short. Frequent wars provided a steady supply of slaves who did most of the work and had no rights.

The rebirth of Western civilization in the Middle Ages included a rebirth in medical knowledge and research. Observation and experimentation flourished in the great universities of the twelfth and thirteenth centuries. Occupational disease received more attention, in general, than work injuries. In 1473, Ellenborg, a German physician, published one of the first known pamphlets on occupational disease. In the sixteenth century, the alchemist Paracelsus wrote extensively about occupational disease in the smelting industry. In 1700, Ramazzini, an Italian physician, published *De Morbis Articisian,* generally regarded as the first integrated treatise on occupational illness and injury. Percival Pott recognized the link between chimney sweeping and scrotal cancer and was a major force in the passage of the Chimney Sweepers Act of 1788 in England. Charles Thackrah, quoted above, wrote an influential treatise on occupational medicine. Despite these advances in knowledge, however, there were few practical advances.

Safety Management Under the Guild System

In the Middle Ages, production was largely controlled by the guild system. Guilds had evolved over centuries as a way of protecting the interests of the trade. They were instrumental in developing and maintaining standards for quality and safety. Another important service was the provision of assistance to ill members and their families—an early form of workers' compensation.

Products were largely handmade. Occupational hazards were less severe than those that were to come with mechanization. In addition, the apprenticeship process ensured that rudimentary safety knowledge was passed on. The master craftspeople could see the value of high quality and uninterrupted production. Large industrial facilities existed, of course, such as the Arsenale, Venice's famous shipbuilding facility, but even here production was largely controlled by the guilds.

Safety Management During the Industrial Revolution

The long decline of the guilds began with the mechanization of industry in the nineteenth century. There was widespread social unrest as machines obsolesced traditional ways of life and left thousands without work. In England, the Luddites were involved in many violent riots, but it was a losing battle. Eventually, the craftspeople gave up and took their places at the machines.

But the births of industrial power and industrial safety were not simultaneous; this was a great tragedy. Mechanized industry created living and working conditions that beggar description. The factory towns were swelled by the steady flow of displaced agrarian workers and peasants. Debasement and social deprivation quickly followed. Manchester, England, grew to 200,000 people yet contained neither a playground nor a park. There was no infrastructure to deliver basic needs such as water and the most primitive sanitation. There were no schools, and living quarters were poor. Deformity and mental illness became commonplace. The death rate tripled.[3]

Factory conditions were treacherous. Factories were little more than shacks with low ceilings, narrow aisles, and little light or ventilation. Rest rooms were unheard of. Two-thirds of the workers were women and children whose typical workday was 12 to 14 hours. Mechanization increased the severity of hazards by orders of magnitude. Yet, machine guarding was unknown. Death and mutilation were commonplace.

The Evolution of Safety and Health Law

As factories conditions came to light, there was public outrage that lead to legislation—a pattern that was to recur several times. The first of a series of English factory laws was passed in 1833. These addressed basic problems such as work hours, minimum working age for children, and basic machine guarding. In 1867, Massachusetts instituted factory inspection and, in 1877, required the guarding of dangerous moving machinery.

Prior to such legislation in the United States, Canada, and the United Kingdom, the only legal recourse was a lawsuit by an injured worker or by the family of a deceased worker. Such lawsuits were usually unsuccessful because of three powerful common law defenses available to the defendant employer.

1. The employee contributed to the cause of the accident (contributory negligence).

2. Another employee contributed to the accident cause (the fellow servant rule).

3. The employee knew of the task-associated hazards involved in the accident before the injury and still agreed to work in the condition for pay (assumption of risk).

These defenses, called the unholy trinity, were given the widest scope in order to ease the employer's burden. Factory owners seldom paid damages. Most felt they were discharging their duty if they provided someone who had lost an eye or a limb with a job pushing a broom. Legislation gradually whittled down these defenses, but it took a long time.

Death and injury rates among workers remained at horrendous levels well into the twentieth century. The historic Pittsburgh Survey completed in Allegheny County, Pennsylvania, in 1909 revealed that there had been 526 fatal industrial accidents in that county alone over a 12-month period. It also revealed that more than 50 percent of surviving widows and children were left with no source of income. The same study estimated that there were 30,000 fatal industrial injuries in the United States in 1909.[4]

The Pittsburgh Survey resulted in the rapid passage of workers' compensation laws in Wisconsin and New Jersey. These laws represented a historic trade-off: The worker gave up his right to sue, but gained guaranteed protection against income loss to industrial injuries and disease. Employers' marginal gain was some measure of predictability in payment for compensation. By 1948, every American state had passed workers' compensation laws.

The regulation of industrial safety in the United States remained a state responsibility until the late 1960s when a movement toward federal primacy began. The quality of state inspectors was seriously questioned, and many other failures were enumerated as well.[5] In 1970, after three years of political posturing and horse trading, the Occupational Safety and Health Act became law. A torrent of workplace safety and health regulation followed.

Safety Management in the Twentieth Century
Role of the Insurance Industry

The most important developments in twentieth-century safety management were not legislative, however. By the mid-1920s the insurance industry had become the driving force for safety. Workers' compensation laws made it imperative that employers purchase insurance to ensure that injured workers were fairly compensated. Insurance companies found it necessary to send inspectors to visit their new policyholders in order to classify risks according to hazards so that proper rates might be charged. This process was already used in fire and property coverage. As inspectors became more competent in assessing risks, they took on an advisory role regarding the reduction of industrial hazards. There were some drawbacks in this situation.

- If it wasn't compensable, it was ignored. In particular, industrial diseases were ignored because they had not been written into workers' compensation laws.

- The most attention was spent on large accounts (which needed it least).

- There was a potential conflict of interest in that insurers were working for the employer and not the employee.

- Insurers were interested in the absence of product liability rather than in inherent product safety (like the rest of society at this time).

Nonetheless, insurers took on the main burden of accident prevention and shouldered it for almost half a century. In doing so, they acquired great insight into the accident process.

Heinrich and Scientific Safety Management

It was an insurance man, Herbert W. Heinrich, who first applied scientific principles to accident prevention. Heinrich based his insights on data from thousands of accidents. He summarized his ideas in *Industrial Accident Prevention*, published in 1931. Dan Petersen said,

> *Perhaps it was because Heinrich proposed a philosophy for safety that his work was so important. Before publication of his book, safety had no organized framework or thinking. It had been a hodgepodge of ideas. Heinrich brought them all together and defined some excellent principles out of previously uncertain practices. . . . As a result, safety progressed markedly after 1931.*[6]

Heinrich's many innovations helped to diffuse the crippling fatalism of both workers and employers toward safety. Prior to Heinrich, an accident was seen as an act of God about which nothing could be done.

Central to Heinrich's system is his domino model of accident causation. Heinrich wrote that "Occurrence of preventable injury is the natural culmination of a series of events or circumstances . . . One is dependent on another and one follows because of another."[7] The model is displayed in Figure 4.3. An accident is the result of one of five factors in a sequence.

1. The injured worker's ancestry and social environment

2. A personal flaw such as violent temper, nervousness, or ignorance

3. An unsafe act and/or mechanical hazard

4. The accident

5. Injury[8]

There are several important features of this model. First, there is a linear sequence of causes: Each domino has only one cause, and that cause is the only cause. The root causes of accidents in this model are the ancestry and social environment of the worker, which lead to

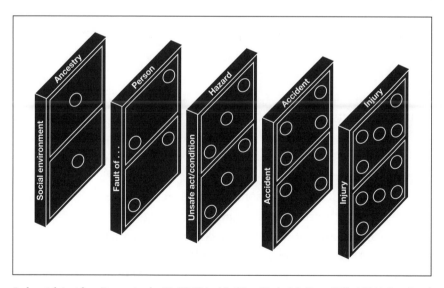

Industrial Accident Prevention by H. W. Heinrich (New York: McGraw-Hill, 1931). Reprinted with permission of The McGraw-Hill Companies.

Figure 4.3. H. W. Heinrich's domino theory of accident causation.

undesirable character traits such as recklessness, bad temper, nervousness, or ignorance. These personal faults lead to unsafe acts and unsafe conditions. Heinrich believed that 88 percent of accidents were due to unsafe acts by workers such as starting up equipment without warning, removal of machine guarding, and so on. He thus mirrored Taylor's belief that workers were essentially untrustworthy and had to be closely supervised.

Accidents, then, happen because a chain reaction occurs and each domino knocks over the domino to the right. How does one control accidents under this model?

Obviously, one can't fix defective people. Heinrich proposed that we focus on the middle domino: unsafe acts/conditions. By removing the middle domino, the accident causation chain is broken and the injury will not occur. Supervisors thus became safety cops, enforcing rules and exhorting workers to work safely.

Weaknesses in Heinrich's Approach

What is the practical effect of Heinrich's approach to accident prevention?

Management avoids responsibility for the system. Rather, responsibility for safety is fixed with the worker, which amounts to blaming the victim. The idea that workers are to blame for accidents is still widely held, just as most managers still believe that workers alone are responsible for poor quality.

Second, focusing on unsafe acts requires heavy supervision and discipline, which are time-consuming and negatively affect morale. Third, the concept of a managed system is absent from Heinrich's approach. Also lacking, therefore, is the concept of continual improvement and upstream control. Finally, the worker is excluded from participation in safety management.

Heinrich and Shewhart Compared

It is interesting to contrast Heinrich's ideas with those of Walter Shewhart, who was a contemporary. In his empirical studies of variation in systems, Shewhart found that

- 85 percent of the causes of error belonged to the system (these were deemed common causes).

- 15 percent of the causes of error originated outside the system and were, therefore, special causes.

Because management controls the system, it follows that

- Management is responsible for 85 percent of the errors.

- The worker is responsible for 15 percent of the errors.

Heinrich, as we have seen, believed precisely the opposite. How can he have been so wrong?

Heinrich reflects the values of his day and its predominant management model: the Taylor system. As an industrial engineer, Heinrich was steeped in the concepts of scientific management. He was also limited by his position as an insurance executive. His job was to serve large industrial accounts. He was not a scientist trying to develop a scientific model of accident causation. Heinrich could hardly have suggested to his clients that management was responsible for the system and therefore for accidents.*

Shewhart, by contrast, was a scientist trying to develop tools with which to improve industrial processes. In his case, the process was the manufacturing of automatic telephone switches. But the underlying principles he discovered were applicable to all systems, including safety.

---

*Frank Bird was similarly constrained early in his career. Later, when Bird left the insurance industry to found the International Loss Control Institute, he directed his attention upstream to the management system itself.

Safety Management Is Rooted in the Taylor System

Safety management in most corporations is based on the traditional safety program—which is rooted in the Taylor system. The history of safety management in the twentieth century corresponds to that of quality management. In both, the application of scientific management principles resulted in tremendous performance improvements. But as the Taylor system became obsolete, performance suffered and now is in a state of crisis. Let us briefly review a series of seminal studies of safety management in the United States in support of this thesis.

In 1989, Veltri surveyed 100 safety professionals in charge of administering safety in their organizations to determine the strategies being used. Veltri examined current safety strategies, perceived needs for the 1990s, and reforms needed to create value for the businesses being served. Veltri identified three distinct strategies.

1. **Reluctant compliers (77 percent).** This group focuses on regulatory compliance and prefers minimal safety investment. It pursues traditional inspection activities, the goal being regulatory compliance rather than accident prevention per se. The safety function's job is to insulate the rest of the organization from compliance problems. This group tends to place blame for compliance demands on government agencies.

2. **Followers (16 percent).** This group approaches safety problems creatively and often employs modern management practices. It uses tools and paradigms typically developed by leaders in their industries. This group recognizes the need to catch up, which it believes is a matter of doing what others have done before.

3. **Leaders (7 percent).** This group employs the best practices in the industry. There is a shared passion for balanced excellence starting in the boardroom and cascading through the organization. There is a relentless focus on the system and on upstream improvement.[9]

During the 1970s, the National Institute of Occupational Safety and Health (NIOSH) studied management and safety program characteristics in high, low, and extremely low accident rate companies. The objective was to determine which factors influence safety performance most strongly. The results were surprising.

> *Commonly prescribed safety practices related to safety committees, safety rules, accident investigations and reporting, and safety promotions were evident in companies with good safety records as well as in those with poor safety records. These factors, therefore, are not differentiating ones.*[10]

In other words, both poor and excellent companies had elements of traditional safety programs: safety staff, committees, inspections, rules, and promotions. The differentiating factors included management commitment and involvement, effective planning, management/labor relations, attitude toward employees, employee/supervisor interactions, and quality of jobs.

Finally, in the late 1970s, Cohen and colleagues conducted a series of studies to identify practices that produced superior and inferior safety results. They focused on two variables: (1) traditional safety elements, and (2) core management competencies. Their results clearly indicated that *traditional safety programs had minimal effects on accident rates.* Their key finding: Companies that effectively managed core business processes produced superior safety results.

We can now answer the questions posed at the beginning of the chapter: What is going on with safety management?

> *Traditional safety continues to be the dominant approach. But traditional safety doesn't work.*

This is not news to the safety profession, which is engaged in a good deal of soul-searching.[11]

Traditional Safety Management

Traditional safety programs

- Focus on injuries, illnesses, and other end-of-the-pipeline measures. A narrow view of accident is taken. Near misses, at-risk behaviors, and other upstream measures are not tracked or understood.

- Tend to be reactive, not preventive.

- Consider safety as a separate function rather than a part of all functions.

- Tend to be project-oriented rather than system-oriented.

- Tend to blame the worker for accidents. Often a Theory X view is taken of workers (based on MacGregor's Theory X and Theory Y: Theory X holds that workers do not like work or responsibility and must be closely supervised).

- Focus on the attitudes of workers. The underlying assumption is that, because workers are responsible for accidents, they must have bad attitudes.

- Rely heavily on promotional campaigns to get people to feel responsible for safety and to give them the "appropriate" attitudes.

- Are based on a top-down model. Management bosses, coaxes, and entices. Workers are rewarded with prizes or are disciplined.

- Place a strong emphasis on rules and close supervision of workers.

Thus, traditional safety management emphasizes discipline and top-down control.

There have been a number of important innovations in safety management since Heinrich, including those of Frank Bird, Dan Petersen, and Edward Weaver, and the system safety techniques developed by the American aerospace and nuclear industries. Unhappily, these have not been widely applied. As we have seen, the predominant approach remains traditional.

I do not wish to be misunderstood. The safety profession and workers all over the world owe much to Heinrich, who was the first to

- Apply scientific principles to the accident process.
- Intuit the link between quality and safety.
- Attempt to quantify the cost of safety.
- Base safety decisions on data.

But the assumptions underlying Heinrich's system are no longer valid. The root causes of accidents are not the ancestry or personal flaws of the worker. Increased supervision and heavy discipline do not improve safety performance. Heinrich's position in the safety profession is analogous to that of Frederick Taylor in industrial management. His innovations have resulted in tremendous gains, but there has been a high cost: The safety profession has largely missed the management revolution represented by Deming, Juran, Drucker, Peters, Senge, and others. Deming threw down the gauntlet with characteristic fervor.

> *Management must feel pain and dissatisfaction with past performance and must have the courage to change. They must break out of line, even to the point of exile among their peers. There must be a burning desire to transform their style of management.*[12]

It's time for safety professionals to take up the challenge.

Notes

1. Allan R. Millet, and Peter Maslowski, *For the Common Defense: A Military History of the United States* (New York: Free Press, 1994), 653, Table A.

2. Larry L. Hanson, "Re-Braining Corporate Health and Safety," *Professional Safety* (Oct. 1995).

3. D. A. Weaver, *Industrial Accident Prevention*, 5th ed., Appendix I by H. W. Heinrich et al, (New York: McGraw-Hill, 1980).

4. Frank E. Bird and George L. Germain, *Practical Loss Control Leadership* (Loganville, Ga.: International Loss Control Institute, March 1986).

5. National Safety Council, *Accident Prevention Manual for Industrial Operations*, 7th ed. (Chicago: National Safety Council, 1974).

6. H. W. Heinrich, Nestor Roos, and Dan Petersen, *Industrial Accident Prevention*, 5th ed. (New York: McGraw-Hill, 1980).

7. H. W. Heinrich, *Industrial Accident Prevention* (New York: McGraw-Hill, 1931).

8. Ibid.

9. Anthony Veltri, "Transforming Safety Strategy and Structure," *Occupational Hazards* (Sept. 1991): 149–152.

10. U.S. Department of Health, Education, and Welfare, *Safety Program Practices in Record-Holding Plants* (Morgantown, W. Va.: National Institute of Occupational Safety and Health, Div. of Safety Research, March 1979).

11. Larry L. Hanson, "Safety Management: A Call for Revolution," *Professional Safety* (March 1993).

12. W. Edwards Deming, *Out of the Crisis* (Cambridge: MIT Center for Advanced Engineering Study, 1982).

--------- CHAPTER 5 ---------

Environmentalism—The First Wave Runs Out of Gas

The sleep of reason breeds monsters.

—John Ruskin

The Roots of Environmentalism

Environmentalism has its roots in the Romantic movement of the early nineteenth century. Romantic poets, painters, and composers rebelled against the rationalism spawned by the scientific revolution. William Wordsworth's sonnet, "The World Is Too Much with Us," expresses the Romantic position (and environmentalist values).

> *The world is too much with us; late and soon,*
> *Getting and spending, we lay waste our powers:*
> *Little we see in Nature that is ours;*
> *We have given our hearts away, a sordid boon!*
> *This Sea that bares her bosom to the moon;*
> *The winds that will be howling at all hours,*
> *And are up-gathered now like sleeping flowers;*
> *For this, for everything, we are out of tune;*
> *It moves us not.—Great God! I'd rather be*
> *A Pagan suckled in a creed outworn;*
> *So might I, standing by this pleasant lea,*

Have glimpses that would make me less forlorn;
Have sight of Proteus* rising from the sea;
Or hear old Triton† blow his wreathed horn.[1]

The Heroic Quest and the Conquest of Nature

Indeed, the sixteenth and seventeenth centuries had witnessed a fracture in the relationship between humanity and nature. The alchemical masters of medieval Europe—including Paracelsus, Kepler, and Copernicus—had practiced a powerful synthesis of science and mysticism. The psychiatrist Carl Jung uncovered many elements of modern psychology in the great alchemical works. Their philosophy of nature was animistic: Nature was a living organism; all living creatures had souls.

The philosopher Rene Descartes' famous aphorism, "I think, therefore I am," expressed the fracture brought on by the Age of Reason. Humanity was no longer perceived to be part of the cosmic web of being. The thinking mind was essentially a disembodied observer, able to soar above nature in a heroic quest for truth. In Cartesian science, nature was considered to be a soulless machine, and animals and plants essentially inanimate. Above all, the mind was separate from nature. To this day, scientists have to pretend that they are rather like disembodied minds. Scientific papers have to be written in an aloof impersonal style, seemingly devoid of emotion. Nobody is ever doing anything. Rather, methods are used, data collected, and conclusions drawn.

The goal of the scientific revolution, in the words of Francis Bacon, its high priest, was "to endeavor to establish the power and dominion of the human race itself over all nature."[2] This was not a

---

*Proteus was a Greek sea god who possessed prophetic powers and the ability to assume different shapes.

†Triton was another sea god, half man and half fish, usually represented as blowing a conch-shell trumpet.

new goal. Human history—ever since fire was tamed, tools were first made, animals and plants first domesticated, and cities first built—has involved humanity's mastery of nature. The hero's conquest of wild nature is an ancient and powerful myth found in most cultures. To wit: the Babylonian story of Marduk defeating the sea monster, Tiamat; the Egyptian god Horus defeating the hippopotamus; the Labors of Hercules; Jason and the Argonauts; St. George killing the dragon.

The scientist thus became an archetypal hero, endowed with superhuman powers and engaged in a great quest. He soars beyond the frontiers of knowledge into the unknown, encounters great obstacles, and brings knowledge and power back to humanity. This metaphor still has great power. Here is *Time* magazine describing Stephen Hawking's picture on the cover of his book *A Brief History of Time:* "Even as he sits helplessly in his wheelchair, his mind seems to soar ever more brilliantly across the vastness of space and time to unlock the secrets of the universe."

But there is another myth of equal and countervailing power.

The Faustian Bargain

The story of Faust, the alchemist whose quest for superhuman knowledge and power leads him to make a pact with the devil, is also as old as humanity. The Greeks called such overweening ambition *hubris* and considered it a tragic flaw. Their plays and myths often deal with heroes who are destroyed by hubris. A relevant example is Prometheus, who arrogantly steals fire from the gods and is condemned to eternal torment. The story of the Garden of Eden is an even earlier incarnation. Adam and Eve are expelled from the garden for eating from the Tree of Knowledge. A more recent retelling is Mary Shelley's *Frankenstein: Or the Modern Prometheus.* But unlike Faust, Frankenstein is not punished by devils; he is destroyed by the monster he created. Frankenstein continues to haunt us, and not only in horror movies. We have created many monsters that threaten to destroy us, not the least of which are our nuclear weapons.

Faust has been the subject of countless plays, poems, and novels. His changing fortunes reflect the spirit of the times.[3] In sixteenth-century incarnations, Faust is carried off to hell by devils. In the progressive nineteenth century, he could no longer be condemned. His restless striving for knowledge was considered to be good, not evil. Thus, at the end of Goethe's *Faust* (1808) he is saved by angels and carried up to heaven. Indeed, one's response to Faust is a reasonable predictor of one's stand on progress and the environment.

The New World

> *There was a time in this fair land when the railroad did not run*
>
> *When the wild majestic mountains stood alone against the sun*
>
> *Long before the white man and long before the wheel*
>
> *When the green dark forest was too silent to be real*
>
> —"Canadian Railroad Trilogy" by Gordon Lightfoot
> ©1967, 1996 Moose Music Inc. Used by permission

Romanticism found expression in the New World during the nineteenth century in the writings of Emerson and his disciple, Thoreau. Like Wordsworth, they expressed a reverential attitude toward nature that was rare in their time. Emerson did not perceive the opening of the West to be a threat to nature. He had a new vision, which is still visionary today: In America, alienated humanity would reassume its place in nature, rather than its war with nature.[4]

Thoreau, by contrast, did perceive a threat to nature. He was not opposed to logging, settling, and cultivation, but he felt they should be practiced in moderation. Thoreau proposed, in vain, that each town in Massachusetts set aside a 500-acre parcel of woodland that would remain forever wild. His most famous book, *Walden,* describes his experiences living in the woods by Walden pond. He built a house, cleared the land, and planted a bean field. In splendid isolation Thoreau came to worship nature.

> *The indescribable innocence and beneficence of Nature—of sun and wind and rain, of summer and winter,—such health, such cheer, they afford forever!* . . . *Shall I not have intelligence with the earth? Am I not partly leaves and vegetable mould myself?*[5]

The myth of the heroic quest and the myth of Faust also found expression in the New World. The opening up of the North American wilderness, a quest of unprecedented scale, continues to be celebrated in movies and novels. Its heroes, from Davy Crocket to Jesse James to Wild Bill Hickock, have entered the popular imagination.[*]

By the 1850s, the railroads had opened up the wilderness. Relentless waves of speculators and settlers moved out into the territories. As they moved west, meat was needed. Buffalo, available by the million, were slaughtered first for meat, then for their hides, and finally for pleasure. During the height of the buffalo hide industry, 1872 to 1874, more than nearly three million buffalo were killed to supply it. By 1880, the unthinkable occurred: The buffalo were nearly gone. By the end of the century, fewer than a thousand buffalo were left; two decades earlier there had been magnificent herds of 30 to 50 million.

The Plains Indians suffered a similar fate. The great plains had to be cleared of the "savages" if America was to achieve its destiny. After the American Civil War ended, the guns were turned westward. Within 25 years, the Plains Indians were annihilated.[6]

The Conservation Movement

The eloquence of Emerson and Thoreau and the excesses that accompanied the opening of the West helped spawn the conservation movement, the forerunner to today's environmentalism. The great

---

[*]Alas, the great Canadian adventurers, including Cartier, Champlain, MacKenzie, and Franklin, are not as well known.

Emersonian lover of nature, John Muir, became a powerful lobbyist for the preservation of large tracts as national parks. The first of many national parks, Yellowstone, was created in 1872. In 1892, Muir founded the Sierra Club, now with more than 500,000 members and a continuing mission to "explore, enjoy and protect the wild places of the earth." The National Audubon Society was founded in 1905, with its mission to protect wildlife. President Theodore Roosevelt, an ardent conservationist, supported legislation protecting "wild places." Conservationist legislation continued to be passed well into the 1930s.

The First Wave of Environmentalism

When did environmentalism per se come into being? There are various interpretations, but in my view, environmentalism began after the first nuclear bomb was dropped on Hiroshima. This cataclysmic event, and the nuclear arms race that followed, illustrated humanity's Faustian bargain with technology in horrific detail. More examples were gradually discovered: smothering smog in large cities, eutrophication of waterways, teratogenic effects of chemicals such as thalidomide, the effects of widespread pesticide use, and so on. Environmentalism gained momentum in the 1960s. Youth, disaffected with the materialism of the day and with the Vietnam War, sought communion with nature much as the Romantics had done 150 years earlier. The first wave of environmentalism crystallized as a political force in 1970, when 20 million Americans came together on the first Earth Day.

Today, public concern for nature continues to grow. At last count, federal environmental regulations comprise 17,000 pages of fine print. Industry and cities spend billions every year on pollution control. Indeed, some industries have adopted stringent codes of practice that exceed regulatory requirements. The chemical industry's Responsible Care* program is an example. International activity has also increased.

---

*Responsible Care comprises a series of environmental codes of practice governing various aspects of the chemical industry. These codes are conditions of membership of the Canadian Chemical Producers Association.

In 1992, leaders from more than 170 countries gathered at Rio de Janeiro for the first Earth Summit. They signed sweeping treaties dealing with global climate and biodiversity.

Successes of the First Wave

The first wave has scored major successes. In the Western developed world, air and water are much cleaner than they were a generation ago. Ambient concentrations of all the major air pollutants—particulates, carbon monoxide, sulfur dioxide, nitrogen oxides, and volatile organics—have declined steadily. Water quality parameters such as phosphorus, dissolved oxygen, and fecal coliform bacteria levels have also steadily improved. Commercial fishing is returning to the Great Lakes and to many smaller lakes and rivers that were barren not long ago. Automobiles are far cleaner to operate; belching smokestacks are rare. Cancer rates are declining for all major cancers with the exception of lung cancer—which is caused predominantly by smoking and is preventable.[7] In summary, the Western developed societies have come together to clean up much of the pollution produced by industry and cities.

The Rise of the Apocalyptics

Such positive statements may come as a surprise. Doom haunts the popular media. Over the past 30 years, society has been besieged by prophecies of impending global disaster.

> • *The battle to feed humanity is over. In the 1970s the world will undergo famines—hundreds of millions of people are going to starve to death in spite of any crash programs embarked upon now.*[7]

> • *The limits to growth on this planet will be reached sometime in the next one hundred years. The most probable result will be a rather sudden and uncontrollable decline in both population and industrial capacity.*[8]

> • *Global warming, ozone depletion, deforestation and overpopulation are the four horsemen of a looming apocalypse.*[9]
>
> • *The threat of new ice age must now stand alongside nuclear war as a likely source of wholesale death and misery for mankind.*[10]

A central characteristic of the first wave has been the rise of the apocalyptics.* Prophets proclaiming imminent calamity are not new in the history of Western culture. In the nineteenth century, for example, various so-called millenarian Christian sects sprang up. These sects believed that Christ's Second Coming was imminent and, with it, the Last Judgment and the destruction of the corrupt world.[11] Neo-Luddites have also found a home among first-wave environmentalists. The most extreme of these, known as deep ecologists, espouse a return to a hunter-gatherer society as humanity's only hope for salvation.

Apocalypse Not

The first wave has turned out to be wrong about many things.[12] The terrible threats predicted in the early days of the environmental movement turned out to be exaggerated. The worldwide famines predicted to occur in the 1970s never happened. Fears that the world's oil and mineral resources would be used up have been refuted by resource gluts and falling prices. Fears that the United States and Europe would cut down all of their forests have been dispelled by increases in forest

---

*It is not my intention here to denigrate the thousands of people who every day donate their time and money to environmental causes. We owe much to their continuing good work and dedication. Rather, I wish to engage the failures of the first wave in a forthright but constructive manner so as to suggest how we might improve.

area. Global warming, despite so many continuing reports, does not appear to be a serious problem. And pesticides caused far less damage to human health and the natural world than Rachel Carson feared when she wrote *Silent Spring* in 1962.*

The Failure of Theory: Malthusianism

A core problem of the first wave of environmentalism has been a failure of theory. From Rachel Carson (*Silent Spring*) to the Club of Rome (*The Limits to Growth*) to Paul Ehrlich (*The Population Bomb*), the leaders of the first wave all operated under a false assumption: Malthusianism.[13]

In 1798, Thomas Malthus, in *An Essay on Population,* argued that humanity was likely to multiply far faster than the food supply. Hence, a portion of humanity would always starve. First-wave environmentalists embraced this notion. They argued that humanity and its technology are heedlessly using up the earth's resources. They liken our situation to that of deer, uncontrolled by predators or disease, overgrazing their pasturage and bringing starvation to all. But they fail to realize that people, unlike deer, can react and adjust their behavior. People can use their intelligence to increase resources and modify their activities to avert Malthusian disasters. In short, if the earth's population was once 1 billion and 10 percent of the earth's oil supplies had been consumed, it is a mistake to assume that 2 billion people would use 20 percent or that 6 billion would use 60 percent. Instead, people have figured out how to use less oil.

---

*Some of these statements may prove to be contentious. Wildavsky's *But Is It True? A Citizen's Guide to Environmental Health and Safety Issues* and *The True State of the Planet,* edited by Ronald Bailey, provide detailed discussions of these issues.

The Pareto Principle—Ignored

The good cannot be held hostage to the perfect.

—Ronald Bailey

The first wave has also tended to ignore the Pareto principle. In terms of many pollutants, the industrialized countries are facing the problem of the last 5 percent. Cleaning up the first 95 percent of many pollutants is comparatively easy and inexpensive, compared to cleaning up the last 5 percent. There is a point of diminishing returns. The issue has arisen repeatedly in the context of contaminated industrial sites, PCB and dioxin contamination, and the incineration of wastes. Is devoting resources to cleaning up the last 5 percent better for the planet than directing those resources to other problems?

The Limits of Command and Control

When the law loses its connection to common sense, no internal system can guide people as to right and wrong.

—Philip K. Howard

Malthus assumed that human behavior would continue unchanged into the future. But the first wave has made an even worse assumption: If behavior does not change on its own, it can only be changed by top-down control.

We have already seen how top-down control contributed to the failure of the nineteenth-century industrial model and the Taylor system. Ironically, first-wave environmentalists have made the same mistake.

The late Aaron Wildavsky has described how, time and again, a headlong rush by regulatory agencies to cure a problem often leads to more harm.[14] For example, the so-called Super-fund program* was designed to facilitate the clean-up of contaminated industrial sites. The

---

*The Super-fund program is operated under the authority of the Comprehensive Environmental Response, Compensation and Litigation Act of 1980 (CERCLA).

program provides that before such land can be used it must be cleaned to extremely high standards. Moreover, any company that purchases such property is liable for previous contamination. The result? Industry moves to virgin fields where it encounters no such costs. Instead of cleaning up one contaminated site, Super-fund effectively creates a second. Over the long term, jobs are lost and the vicious circle of inner city decay continues. Thus, the effect of the Super-fund has been perverse: more pollution and fewer jobs.

A second example of the perverse effects of regulation was the removal of asbestos from schools and public buildings in America. This was mandated by the EPA, even though the asbestos posed no significant threat to health, at a cost exceeding $150 billion. The exposure of building occupants actually increased in some cases because the removal disturbed the asbestos. But the most serious exposure was incurred by thousands of removal workers. Many fear that there will be another wave of asbestos-related disease among them.[15]

The most serious cases of environmental degradation have proved hard to fix by law. The wretched state of global fisheries is a case in point. Overfishing results from the all-too-familiar problem known as the "tragedy of the commons."[16] The analogy to overfishing is overgrazing of lands held in common. When land is open to anyone who wants to use it, the tragedy of the commons almost inevitably results. In the case of commonly held grazing land, herders have every incentive to increase the number of cows grazing on the commons. If they do not, their neighbors will, and will thus reap the benefits of raising an additional cow. By contrast, there is no incentive to limit the number of cows grazing on the commons. Overgrazing and the eventual destruction of the common pasture inevitably result. This is what has happened to many of the world's fisheries. If everybody owns it, nobody owns it—and nobody looks after it.

First-wave environmentalists have failed to realize that the problem lies in the commons, not in the herders. They usually want to regulate the herders instead of abolishing the commons. The way to avoid the tragedy of the commons is privatizing resource ownership. If individual herders can fence in portions of the commons and secure ownership rights and responsibilities, their incentives to protect the land dramatically increase.

The overemphasis on regulation is part of a broader trend in America and Canada. According to Philip Howard, the idea of law has been ridiculously oversold. We are drowning in law, legality, and bureaucratic process.[17] As noted earlier, environmental regulations in the United States now comprise more than 17,000 pages of fine print. The Occupational Safety and Health Act comprises more than 4000 regulations. Who can possibly know, much less comply with, these laws?

Howard contends that the law has abandoned common sense. Regulation has become so complex that it is "beyond the understanding of all but a handful of mandarins."[18] As a result, Howard argues, there is a resentment of government and a collective powerlessness. This passivity in the face of the law corrodes individual responsibility and subverts democracy.[19]

The Faustian Bargain Again

It is time to declare my biases (if I have not already done so). I believe we must accept the Faustian bargain inherent in the quest for knowledge. That is not to say that we should pursue progress and technology blindly. We should go forth with humility and a clear understanding of the bargains we strike. The reintegration of humanity into the web of being would also be a welcome philosophical development. As we shall see, the cutting edge in environmental management—industrial ecology—anticipates this development.

But environmental improvement depends on economic progress. Richer is cleaner and safer, regardless of what deep ecologists say. This is why impoverished people in Bangladesh die by the thousands when cyclones strike their villages, while few Americans die when hurricanes hit Florida. Better roads, housing, medical facilities, and emergency response measures made possible by American wealth make it far easier to weather storms in the United States. In the developed countries, air and water pollution have been significantly cut during the last 25 years. Clearly, as societies become wealthier, they use some of their wealth for environmental improvement. Impoverished societies have few resources to devote to protecting and cleaning up their natural environments.

Moreover, the preservation of natural resources and the expansion of human ones can be complementary. If poor countries do not adopt modern high-yield agriculture, for example, their impoverished farmers will be forced to level millions of square miles of wilderness. Agricultural intensification is essential to avoid famine and the plowdown of wildlife habitat. Currently, more than 75 percent of the land on every continent except Europe is still available for wildlife. This undeveloped land in developing countries is vital to conserving biodiversity over the long term.

Modern forestry also helps preserve wildlife habitat. Although nearly 75 percent of total industrial wood production comes from industrialized countries in the Northern Hemisphere, the forests of this region are expanding. With modern technology, the world's current industrial wood consumption requirements could be produced on only 5 percent of the world's total current forestland.[20]

Environmentalism—The Second Wave

The command-and-control approach central to the first wave of environmentalism has reached its limits. More environmental regulations are unlikely to benefit nature and may even be counterproductive. We must shift our focus from the end of the pipe to the system itself. We must apply the quality approach to environmental management. Cutting-edge companies have been doing this for some time. The term *total quality environmental management* (TQEM) has been coined to describe the new approach.[21,22]

The key difference between the second wave and the first wave will be its approach to solving problems: not by fiat, but by freely available and accurate information; not by doomsaying, but by applying the PDCA cycle (current status, desired status, countermeasures) to pollution prevention.

Summary

Environmentalism has its roots in the Romantic movement of the early nineteenth century. The Romantics rebelled against the materialism of the Age of Reason. They rejected Descartes' mechanistic view of nature and instead sought spiritual communion with nature. The Romantic ideals found expression in the New World in the writings of Emerson and Thoreau. Their ideas informed the conservation movement that came into being in the latter half of the century and is the precursor to modern environmentalism.

Two competing archetypal myths inform progress and environmentalism: the heroic quest for knowledge and the story of Faust. Understanding these myths clarifies much of the present discourse on environmental matters.

Environmentalism per se came into being after Hiroshima, which illustrated the Faustian bargain with technology in horrific detail. The first wave of environmentalism crystallized as a political force in 1970 with the first Earth Day, and it has scored major successes. In the Western developed world, air and water are much cleaner than they were a generation ago. The first wave has been characterized by the rise of apocalyptics who predict impending global disaster; however, the apocalyptics have usually turned out to be wrong.

Core problems with the first wave of environmentalism include

- The failure of theory: Malthusianism
- Ignoring the Pareto principle
- The emphasis on top-down control

The emphasis on top-down control has spawned a mountain of regulation that no one can understand. The most serious cases of environmental degradation have proven hard to fix by law. Environmental law, like regulatory law in general, has abandoned common sense, which has created resentment and a sense of powerlessness in the populace.

I believe that we must accept the Faustian bargain inherent in the quest for knowledge, but that we should pursue knowledge with humility and a clear understanding of the bargains we strike. Environmental improvement depends on economic progress. Moreover, the preservation of natural resources and the expansion of human ones can be complementary.

The second wave of environmentalism requires a shift in focus from the end of the pipe to the system itself. The quality triad—leadership, measurement, and participation—will drive the second wave. Cutting-edge companies are already applying these ideas.

Notes

1. William Wordsworth, *Selected Wordsworth,* ed. by George W. Meyer (Arlington Heights, Ill.: Harlan Davidson, 1950).

2. Rupert Sheldrake, *The Rebirth of Nature* (Rochester, Vt.: Park Street Press, 1991), 40.

3. Ibid., 39.

4. Ralph Waldo Emerson, *Selected Essays* (Harmondsworth, England: Penguin Books, 1985).

5. Henry David Thoreau, *Walden* (London: Penguin Books, 1988), 314–315.

6. Sheldrake, *The Rebirth of Nature.*

7. Paul Ehrlich, *The Population Bomb* (New York: Sierra Club–Ballantine, 1968).

8. Donella Meadows et al., *The Limits to Growth* (New York: New American Library, 1972).

9. Michael Oppenheimer, "From Red Menace to Green Threat." *The New York Times Magazine,* 27 March, 1990. (Oppenheimer holds the Barbra Streisand research chair at the Environmental Defense Fund.)

10. Nigel Calder, "In the Grips of a New Ice Age?" *International Wildlife* (June 1975).

11. Ronald Bailey, *Eco Scam—The False Prophets of Environmental Apocalypse* (New York: St. Martin's Press, 1993).

12. Aaron Wildavsky, *But Is It True? A Citizen's Guide to Environmental Apocalypse* (New York: St. Martin's Press, 1993).

13. Ronald Bailey, *The True State of the Planet* (New York: Free Press, 1995).

14. Aaron Wildavsky, "The Secret of Safety Lies in Danger," in *The Constitution and the Regulation of Society,* ed. Gary C. Bryner and Dennis L. Thompson (Provo, Ut.: Brigham Young University, 1988).

15. Wildavsky, *But Is It True?*

16. Peter Senge, *The Fifth Discipline* (New York: Currency Doubleday, 1990).

17. Philip K. Howard, *The Death of Common Sense: How Law Is Suffocating America* (New York: Warner Books, 1996).

18. Bayless Manning, as quoted in Howard, *The Death of Common Sense.*

19. Howard, *The Death of Common Sense.*

20. Bailey, *The True State of the Planet.*

21. Global Environmental Management Initiative, *TQEM: The Primer* (Washington, D.C.: GEMI, 1994).

22. *Total Quality Environmental Management* (New York: John Wiley & Sons, 1991).

CHAPTER 6

Systems

The total quality system is the foundation of total quality control.

—Armand V. Feigenbaum

The systems approach has its roots in general systems theory, which was developed to support the computer revolution. The electrification of data resulted in a tremendous increase in both the speed and quantity of information flow. It has obsolesced the linear mode of thinking characteristic of the Taylor system. The speedup of information forced an integration of sources of data and knowledge that used to be compartmentalized.[1]

Systems thinking with respect to quality has its roots in Walter Shewhart's work at Bell Laboratories in the 1920s. As we have seen, the concepts such as focusing on the system, upstream prevention, and continual improvement are central to the quality approach.

The Systems Approach to Management

The systems approach has been adapted to solve broad management problems. A *management system* may be defined as an orderly set of components that serves to accomplish one or more goals of the organization. A system's specific goal may be to facilitate the flow of information, improve quality, minimize losses due to accidents and injuries, or reduce environmental impacts. The underlying goal of all management systems, however, is to reduce entropy.

In the absence of a systems approach, companies may be overwhelmed by the increasing complexity of the business environment. The components of this complexity are technological, organizational, and marketing. Safety and environment bring the added dimension of legal liability. Root causes of this increasing complexity include the globalization of business, relentless consumer demand for higher quality at lower costs, and the information explosion brought about by the computer revolution.

According to Feigenbaum, a clear and well-structured system "identifies, documents, and coordinates, and maintains all the key activities needed to assure the necessary quality actions throughout all the relevant company and plant operations."[2] A system channels, filters, and organizes complexity. It provides order, structure, and constancy of purpose. A system thus acts as an anchor in the sea of complexity. This is not to say that systems should be static. Effective systems remain flexible through their auditing and change control elements.

Systems are also a strong integrative force. Both the factory system and the Taylor system called for improvement by a specialized division of effort and a vertical or top-down assignment of responsibilities.

This encourages disintegration of effort. Each division of a company does its work and then throws the product over the wall to the next sequential step. By contrast, the systems approach calls for improvement by the integration of effort and the horizontal assignment of responsibilities.

The Systems Challenge

The systems challenge is substantial.

- Quality, safety, and environment systems require a tremendous increase in the speed and volume of information.

- Quality, safety, and environment are demanding concepts to define and structure in complex organizations.

- Managerial skills needed to operate these systems are not widely practiced; both the hard data management skills and the softer skills (those associated with integrating new forms of information into decision making) are required.

- The system must overcome individual prejudices and organizational barriers (both structural and cultural).

The systems challenge was underestimated during the second wave of SPC in the 1980s. This was a broad movement to train company personnel in SPC and other quality tools spurred by the CBS documentary "If Japan Can, Why Can't We?" The assumption was that the Japanese success in quality was due largely to statistical methods. Such training provided a useful set of tools, but it was premature. SPC was introduced in a systems vacuum, before companies had defined their quality goals and strategies. The quality tools could not be integrated into decision making. Disappointment was widespread.[3]

System Structure

> *Behavior follows structure.*
>
> —Peter Senge

System structure, as Senge suggests, will strongly affect activity. The hallmarks of a good structure are clarity and simplicity. Accordingly, let me strongly advocate the structure displayed in Figure 6.1.* It comprises the following elements:

- A *policy* statement that describes the intent, values, and beliefs the system is designed to promote

- A set of *principles* that further develop the intent and values of the policy statement

---

*I am indebted to my friend and colleague, Dr. Peter Strahlendorf, for this figure and for many fruitful discussions on the ideas outlined in this section. Indeed, many are his own.

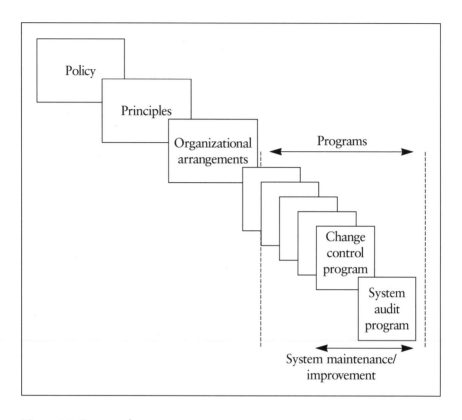

Figure 6.1. Suggested system structure.

- A set of *organizational arrangements* that outline the responsibility, authority, and accountability of relevant positions in the organizational hierarchy for system outcomes
- A set of interlocking *programs,* each of which addresses specific problems that must be solved to achieve the policy
- Special programs that ensure *system maintenance and improvement,* including change control, document control, and audit programs

Let me declare a bias. I find this blueprint to be clearer and simpler than those described in ISO 9000 and ISO 14000. A system thus structured will meet or exceed the requirements of these standards. In the following sections, I will describe its components and will comment on their congruence with the ISO standards.

Policy

A *policy* may be defined as a guide to action, expressing important values or beliefs, that should be followed by individuals in order to attain stated goals and to provide consistency of decisions. The system policy statement should simply and clearly express the company's core values and beliefs and its general intent, goals, and objectives.

The intent, goals, and objectives are related but correspond to different time scales. Intent addresses the long term; goals address the midterm (say, two to five years) and objectives; the short term, usually one year. Objectives cascade into goals that, in turn, cascade into the long-term intent.

The intent, goals, and objectives expressed in the policy should be SMART (simple, measurable, achievable, reasonable, and trackable). This is especially important with the environmental policy, which may have broad strategic implications. The congruence between a firm's environmental policy and its environmental performance may be closely scrutinized by many diverse and critical audiences. These include employees, shareholders, creditors, regulators, environmental groups, and the general public. Setting a goal that is too complex to understand or one that is not measurable or trackable is likely to create cynicism and mistrust. Selecting one that is not achievable could lead to criticism and bad publicity. These concerns also apply to quality and safety system policies. Quality and safety performance, however, are not usually scrutinized by as many audiences. These questions should be addressed during the system planning phase.

In addition, the system policy statement should

- Fit on one page to facilitate its wide distribution (for example, the wall of the lobby, corporate reports, and training and policy manuals).

- Identify quality, safety, or environment as management responsibilities, which nonetheless require the participation of everyone in the company.

- Indicate how the policy will be implemented.

- Indicate how its implementation will be monitored.

- Indicate which senior person is responsible for its implementation.

- Be signed by the highest-ranking official and possibly by the union representative.

- Be dated (the policy should be updated annually so maintain its relevance and rekindle commitment).

Good policy statements manage to say a great deal in a short space.

A policy generated thereby will satisfy ISO 9000 and ISO 14000 with only minor adjustments. ISO 14000 requires that the policy express the principles of continuous improvement and pollution prevention, as well as a commitment to comply with the law. ISO 14000's four-part planning process (described later in this chapter) is consistent with the planning method described here.

Principles

Principles further develop the values, beliefs, and intent expressed in the policy and help keep it to one page. In the principles section of an environmental management system (EMS), for example, you might wish refer to industry codes of practice or to international protocols—such as those arising from the Rio Summit. In the principles section of a safety management system (SMS), you might affirm the importance of cooperation between management and the union, or the right of each worker to a safe workplace. Or you might affirm the commitment to sound ergonomic design. The principles section of a quality management system (QMS) could describe the intent, nature, and frequency of quality audits and so on.

Neither ISO 9000 nor ISO 14000 have analogues to the principles section, but would benefit if they did. Abstract concepts such as *sustainable development*, widely cited in environmental policy statements, require a good deal of explanation best suited to the principles section.

Organizational Arrangements

Organizational arrangements set out the responsibility, authority, and accountability for system activities for all relevant positions in the organizational hierarchy. For safety and environmental systems, legislated responsibilities should also be defined. As a general rule, the more senior the position, the greater the authority, and the greater the responsibility and accountability.

Organizational arrangements should express in clear and simple terms what each person is supposed to do (responsibility) and how performances will be assessed (accountability). Systems frequently founder on the latter. Typically, a rogue manager, otherwise productive, blatantly ignores his or her responsibilities and is given a light tap on the wrist. A clear message is delivered: "We are not really serious about this." Another common breakdown: Responsibility for system outcomes is given to those who lack the authority to change the system. This is an abdication of leadership. System outcomes belong to senior management, and the organizational arrangements must express this.

Organizational arrangements as described here are consistent with the requirements of ISO 9000 and ISO 14000, with an important exception: ISO 9000 and ISO 14000 do not require that you show how management will be held accountable for system outcomes—a significant weakness.

Programs

A *program* may be defined as a sequence of steps or activities required to accomplish objectives under a management system. The programs are the heart of the system, the means by which you will fulfill the policy. In the absence of clearly defined programs, the system is likely to be a general statement of good intentions, procedural statements, and documentation needed.

Both ISO 9000 and ISO 14000 are somewhat weak in this regard. Neither requires a detailed description of the program content of the QMS or EMS—what you are supposed to be *doing* to improve performance. It is not enough, in my view, to say that you are going to investigate all spills. You must also define *how* you are going to do it, *who* will be responsible, *when* reports and follow-up checks will be due, and so on.

System Maintenance and Improvement

The requirement to maintain and improve the system is congruent with ISO 9000 and ISO 14000.

Program Development

Each program should be defined by a clear and simple flowchart. SMART goals should be set. Organizational responsibilities of relevant positions should be defined and documented. Program effectiveness should be assessed using the PDCA cycle. This requires a clear understanding of the current status based on extensive observation and discussion. The desired status should be determined in consultation with those who will be affected by the program. Periodic checkpoints should be established. Countermeasures should be prepared when progress is below par.

Safety Management System

A loss causation model such as Bird's, displayed in Figure 6.2, should be used to rationalize the programs in an SMS. In Bird's model, a loss is preceded by *contact* with an energy source, and he characterizes

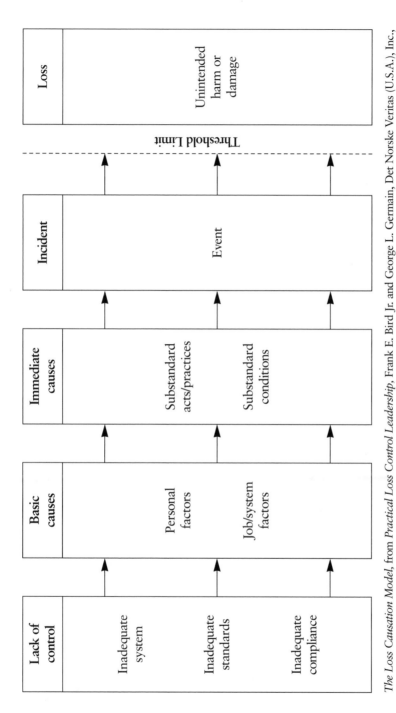

The Loss Causation Model, from *Practical Loss Control Leadership*, Frank E. Bird Jr. and George L. Germain, Det Norske Veritas (U.S.A.), Inc., August 1996, 7. Used with permission.

Figure 6.2. Bird's loss causation model.

programs as precontact, contact, and postcontact.[4] Here are examples of each type of program.

- Precontact: Training, inspections, preventive maintenance
- Contact: Personal protective equipment
- Postcontact: Emergency response, first aid

Environmental Management System

ISO 14000 includes the following four-part planning process, which may be used to rationalize the program mix.

1. *Environmental aspects.* Identify the environmental aspects of the organization' activities. From this list, determine the most important potential environmental impacts.

2. *Legal and other requirements.* Identify the relevant legal requirements, industry codes of practice, best available technology economically achievable (BATEA), and so on.

3. *Objectives and targets.* Set environmental objectives and targets in accord with the environmental impacts and with legal and other requirements identified.

4. *Environmental management programs.* Prepare a program mix to achieve the objectives and targets.

Prioritization

You should develop the most important programs first. Programs in an EMS or SMS can be prioritized as follows (in descending order).

Core programs. Core programs either

- Directly monitor and control the more severe environmental impacts and threats to safety and health.
- Address the general intent, goals, and objectives of the system as outlined in the policy.
- Are explicitly required by law.

Examples of core programs in an EMS include leadership, emergency response, and emissions monitoring.

Programs that support core programs. For example, the training program supports each of the core programs detailed previously.

Programs that enhance system performance. The change control program helps keep the system relevant. The audit program assesses how well the system is working. The document control program facilitates the administration of the system.

Programs that go beyond the core goals of the system. For example, health promotion or off-the-job safety programs might go beyond the core goals of an SMS.

Remember Pareto

> *Less is more.*
>
> —Robert Browning

The Pareto principle should guide program development. A system comprising eight programs that work is preferable to one comprising 40 programs that are ignored. Do not try to solve every problem; rather, focus on the big ones.

Nemawashi

The Japanese principle of *nemawashi* should be applied to facilitate the implementation of programs. This elegant term literally means "digging around the roots" and refers to the careful process inherent in the safe transplant of a tree or plant. For the tree to thrive you must prepare the soil and you must transplant carefully. Nemawashi is hard work. It entails discussions with everyone who will be affected by the program and comprises the following steps.

- Determine the need.
- Establish rapport.

- Grasp the current status.

- Offer a proposal.

- Negotiate an agreement.

- Check for consensus.

In the West, important decisions are usually made at formal meetings. In Japan, by contrast, the most important discussions are held at an earlier stage. Nemawashi entails informal discussions over lunch or a cup of coffee. By the time the meeting occurs, the proposal is acceptable to all—even those who originally might have opposed it.

System Management Activities

Managing an effective system requires the following activities:

- System administration

- System economics

- System measurement

System administration entails coordinating system work across the organization. For example, life cycle analysis, a core EMS activity, requires organizing and controlling input from purchasing, design, production, and marketing. The nemawashi principle, essential at the program development stage, also facilitates such cross-functional activities.

System economics is the measurement and control of system costs. There is now a substantial literature on quality costing. However, such is not the case for safety and environment costs, which continue to be lost in overhead accounts. They are rarely affixed to their source. The accounting profession is working to address these issues.[5,6] In the interim, full cost accounting, which is discussed in chapter 11, can be used to estimate these costs.

System measurement is the process of evaluating the effectiveness of the system. System measurement covers both system outcomes (such as customer satisfaction) and the process by which the results were obtained. (Did the system work as designed?) The former requires surveys of customer satisfaction (both internal and external); the latter requires system audits. As always, use the PDCA cycle to organize thinking.

The Paper Wall

Excessive paperwork is a common cause of system failure. Most of our private and public organizations are designed to handle a much slower and smaller flow of data in paper form. But management systems require a tremendous speedup and increase in the flow of information. In fact, the hallmark of an effective system is the right information, in the right quantity, that flows to the right person at the right time, and allows him or her to make the right decision.

Paper-based systems have difficulty satisfying these criteria. ISO 9000, for example, can be a paperwork nightmare.[7] There continues to be a mismatch between newer forms of information processing and institutions designed for older forms.[8]

A Word on Leadership

Leadership is the wind that fills the sail. In its absence, even the most adroitly structured system will hang limp and useless. Although it is beyond the scope of this book to fully explore the nature of leadership, let me simply summarize what I have seen. The best leaders I have known share the following qualities.

- Strong values and a strong sense of self
- The ability to articulate a vision based on their values
- The ability to inspire others to achieve the vision
- A selfless desire to serve the common good

Leadership is the great equalizer. It can override not only the paper wall but also inadequacies of structure, documentation, or training. Indeed, some industry leaders in quality and environment would fail to meet ISO 9000 or ISO 14000. They excel because of leadership. Leadership permeates their activities from the shop floor to the boardroom.

Summary

In this chapter, the systems approach to management has been described. In the absence of a systems approach, a company may be overwhelmed by the growing complexity of the business environment. The systems challenge with respect to quality, safety, and environment is substantial.

A system structure comprising five components is described. The structure is clear and simple, and congruent with the requirements of ISO 9000 and ISO 14000. System management activities including system administration, economics, and measurement are described. Finally, the nature of leadership and its overriding importance to system success is emphasized.

Notes

1. Robert K. Logan, *The Fifth Language: Learning a Living in the Computer Age* (Toronto: Stoddard, 1995).

2. Armand V. Feigenbaum, *Total Quality Control,* 3rd ed. rev. (New York: McGraw-Hill, 1991) 77.

3. J. M. Juran, *A History of Managing for Quality* (Milwaukee: ASQC Quality Press, 1995), 585.

4. Frank E. Bird and George L. Germain, *Practical Loss Control Leadership* (Loganville, Ga.: International Loss Control Institute, March 1986), 26.

5. Canadian Institute of Chartered Accountants, *Environmental Costs and Liabilities: Accounting and Financial Reporting Issues* (Toronto: CICA, 1993).

6. Price Waterhouse, *Accounting for Environmental Compliance: Crossroads of GAAP, Engineering, and Government* (1992).

7. Perry L. Johnson, *ISO 9000: Meeting the New Standards* (New York: McGraw-Hill, 1993).

8. Logan, *The Fifth Language*, 214.

CHAPTER 7

Total Safety and Environmental Management

The way to prevent firefighting is not to have fires.

—Philip Crosby

Total Safety and Environmental Management Defined

Let me define a new term. *Total safety and environmental management* (TSEM) is all employees working together through a clearly defined system to continually improve safety and environmental performance through statistical tracking of objectively defined upstream measures of performance.

TSEM entails applying the quality approach to safety and environmental management.

The Goals and Methods of TSEM

The goals of TSEM are to

- Protect and enhance employee safety and health.

- Reduce costs due to employee injury and illness.

- Reduce variation in work procedures, materials, and equipment.

- Minimize the environmental impacts of an organization's activities.

- Reduce waste in the form of the following:
  —Emissions to the air, water, soil, or ground water
  —Inefficient use of energy, raw materials and other resources
  —Inefficient use of land

The methods of TSEM include

- The quality toolbox (chapter 4)
- Behavior-based tools (chapter 8)
- System safety techniques*
- The techniques of industrial ecology (chapter 11), in particular
  —Life cycle analysis
  —Design for environment
- The management system (chapter 6)

Let us begin our description of TSEM by discussing the synergy between safety and environmental management.

Synergy Between Safety and Environment

To this point, we have emphasized the potential synergy between quality and safety and between quality and environment, because they are not intuitively obvious. The strong synergy between safety and environment, however, is fairly obvious. The line separating inside the plant from outside the plant is artificial. Many safety issues are also environmental issues. Chemical, physical, and biological contaminants inside the plant often manage to exit the plant. Indeed, safety

---

*System safety comprises a powerful set of tools developed by the U.S. aerospace and nuclear industries to increase the reliability of complex systems. Many system safety tools, including fault tree analysis and failure mode and effect analysis, are now part of the quality toolbox. A detailed discussion of system safety is beyond the scope of this book. The interested reader is referred to *System Safety 2000* by Joe Robertson (New York: Van Nostrand Reinhold, 1994), which is a good introduction.

and environmental practices overlap broadly. Many organizations structure accordingly and group these functions. Often a given professional will have responsibility for both. The corresponding management systems also intersect in important ways.

Common Programs

Examples of common programs include

- *Chemical control* including purchasing, tracking of material safety data sheets, labeling, storage, and transportation
- *Inspection* including inspections of the general plant, critical parts, special systems (fire and other emergency response systems), maintenance
- *Incident investigation* including root cause analysis, countermeasures, follow-up, and incident data analysis
- *Emergency response* to fire, spill response, and disasters
- *Training and education* related to common programs
- *Product liability* including system safety techniques used to minimize product liabilities (the precursors to life cycle analysis)
- *Auditing*

Crossover Programs

Some safety programs have crossover potential (to borrow an expression from the music industry). These include task analysis and behavior-based approaches. Task analysis is a system safety technique whereby a given task is broken down to its constituent parts. The potential safety hazards associated with each part are identified. Critical tasks, those entailing substantial risk, are identified and modified. A common by-product is increased efficiency and quality. Task analysis can also be used to identify critical tasks with respect to the environment and to control associated risks.

Behavior-based techniques, discussed in chapter 8, are applicable to environmental hazards. Essentially, these entail compiling and tracking an inventory of critical behaviors; that is, behaviors that are the common pathway to accidents (or environmental incidents). SPC applied to data generated thereby can provide an early warning of risk in the system.

Differences Between the SMS and the EMS
Strategic

Safety and environment differ in that the latter is taking on broad strategic significance. A firm's environmental performance may be closely scrutinized by many diverse and critical audiences, including employees, shareholders, creditors, regulators, environmental groups, and the general public. Their responses could significantly affect share prices. Environmental performance can also have trade implications. Firms can find themselves shut out of certain markets on the basis of environmental performance.

Safety performance is also scrutinized by employees and regulators, but it is generally less significant to shareholders, creditors, and the public. Safety performance rarely affects share prices and generally has few trade implications.

Structural

There also structural differences between the SMS and the EMS. Core programs can differ substantially. Ergonomics, a key safety program issue, is not an environmental issue. Occupational health programs such as primary health care, medical surveillance, and health promotion have no environmental analogue. Monitoring of in-plant and out-of-plant emissions differs significantly. Public relations and communication, a central environmental program, is of lesser importance to safety management. Design for environment, a core environmental activity in cutting-edge companies, has a comparatively minor worker safety component. Moreover, some environmental impacts such as solid and liquid waste management; the use of energy, raw materials, and other resources; and inefficient land use have minimal safety implications.

Let us now illustrate the principles that inform TSEM by applying Deming's 14 points to safety and environment.*

Principles of TSEM—Deming Revisited

Create constancy of purpose—focus on the long term. Create constancy of purpose toward minimizing injury and illness in employees and environment degradation.[1] To remain competitive and profitable in the twenty-first century, companies must reduce the impact of their operations on employees and on the environment.

Senior management must create constancy of purpose in safety and environment by

- Providing the resources for the establishment of safety and environmental management systems

- Becoming actively involved in these systems

- Providing long-term goals in safety and environment and holding the company and its management accountable for those goals

Protecting the health and safety of employees and the natural environment must be core values of the company, on par with productivity. There is no more powerful message to employees than "You are our most important resource. The future and well-being of our company depend on your well-being. We will do all we can to protect it."

Adopt the new philosophy. Recognize that we are living in a new environmental age. The hydrosphere, atmosphere, and oceans are a global

---

*I have chosen Deming's work to illustrate the application of quality principles to safety and environment. The work of other quality senseis may also be used, to wit, Crosby's four absolutes, Feigenbaum's fundamentals, or Juran's quality triad. The criteria on which the Malcolm Baldrige National Quality Award are based are another possibility.

commons. Injure these systems for personal gain and, eventually, they will injure us all. We can no longer afford to ignore the effects of industrial metabolism on natural systems. Human activities now have the potential to affect the global budgets* of elements essential to life. The assumptions of the Industrial Revolution—unlimited resources and unlimited capacity of the planet to absorb waste—are not longer tenable.

Build in safety and environmental protection. Cease dependence on downstream safety measures such as accident frequency and severity rates. Cease dependence on measurement and treatment of waste streams. Require, instead, statistical evidence that safety and environmental protection are built into the process. Develop leading indicators of safety and environmental performance and act to eliminate risks in the system. Reduce variation in the system by standardizing operating procedures with tools such as task analysis. Select employees for safety and environmental performance. Continually seek less toxic materials and less hazardous processes. Use the techniques of life cycle analysis and design for environment to reduce environmental impacts.

End the practice of awarding business on the basis of price. Instead, award business on the basis of meaningful measures of quality, cost, and environmental impact. The customer expects environmental stewardship and is willing to pay for it. It is no longer acceptable to do business with rogue firms that degrade the environment. The computer revolution ensures that environmentally sensitive practices will prevail.

Continually improve the system. Find problems and fix them. It is management's job to work continually on the system. Reduce

---

*In ecology, a budget analysis is the process of measuring or estimating the flows of a given material into or out of a reservoir and checking the overall balance by measuring the amount present in the reservoir.

dependence on downstream indicators, which trap the organization in hindsight. Focus on upstream measures and manage the system.

Implement modern methods of training. Train managers and employees in the relevant principles of safety and environmental protection. Give them the tools to track upstream results and encourage them to solve problems proactively. Statistical literacy is a condition for being a manager.

Drive out fear. Drive out fear and stop blaming employees for injuries and environmental incidents. Management controls 85 percent of the system, and these belong to that system. Punishing employees for safety and environment creates resentment, whereas risk taking occurs because the system favors it.

Eliminate slogans, exhortations, and targets. Stop demanding higher levels of safety and environmental performance from employees without providing the tools to meet them. Hectoring employees for zero injuries or zero discharges is counterproductive. At best, posters, pep rallies, and exhortations show that management is out of touch. At worst, they show that management wants to pass the buck for safety and environment.

Remove barriers to pride of workmanship—involve the workforce. The experience and skill of employees is an untapped resource in safety and environmental management. Employees have the greatest stake in safety, yet they are rarely invited to participate in safety efforts. Most employees want to work for environmentally responsible firms. Moreover, safety and environment are values on which the workplace parties agree. Let them be bridges between management and labor.

Institute modern methods of supervision. The responsibility of the supervisor must be enlarged to include safety and environment. The supervisors (and managers) of the twenty-first century must be "multiball jugglers."[2] They achieve excellence across a broad range of values through the development of vigorous teams.

Break down barriers between departments. Solving safety and environment problems requires a multidisciplinary approach. The relevant departments including purchasing, design, engineering, and production must work together to anticipate and eliminate problems. Quality, safety, and environment professionals must collaborate to reduce variation in the system.

Put everyone in the company to work on the transformation. Create a structure in which each employee can contribute to excellence. Production is not enough. The world-class organization must achieve "balanced excellence"[3] in production, profitability, safety, and environment. These are not mutually exclusive. The world-class organization strives to continually reduce variation in its processes. Improvements in safety and environmental performance inevitably lead to improvements in quality, and vice versa.

Summary

In this chapter a new term, total safety and environmental management, has been defined. TSEM entails applying the quality approach to safety and environment. The goals and core tools of TSEM were outlined. The strong synergy between safety and environmental management was discussed—as well as some of their differences. The principles of TSEM were illustrated by expressing Deming's points to safety and environment.

Notes

1. W. Edwards Deming, *Out of the Crisis* (Cambridge: MIT Center for Advanced Engineering Study, 1986).

2. J. M. Stewart, "The Multi-Ball Juggler," *Business Quarterly* (Western Business School, The University of Western Ontario) (winter 1993).

3. Ibid.

CHAPTER 8

Behavior Observation—The Key to Upstream Measurement in Safety

Why would a worker choose to risk his life?

—Manager at a large utility

Why indeed? I was asked this question after a near miss involving 13,800 volts. The worker, a skilled veteran, had taken numerous shortcuts in an electrical switching operation and should have been killed. His manager could not conceive why he would behave this way, yet workers take such risks every day.

Let me declare, up front, that some workers do so because they are knuckleheads—but they are in the minority. Most workers take risks because the system encourages and rewards risk taking. In this chapter we will discuss behavior-based safety (BBS), a core element in the safety management system. BBS, the key to upstream measurement and continuous improvement in safety, helps to explain why workers take risks.

The Achilles' heel of traditional safety management has always been the absence of meaningful upstream measures. Upstream measures are especially important in safety because they provide an early warning of accidents. As we have seen, current performance measures amount to a body count.

The accident ratio pyramid displayed in Figure 8.1 shows the relationship between serious injury accidents, minor injury accidents,

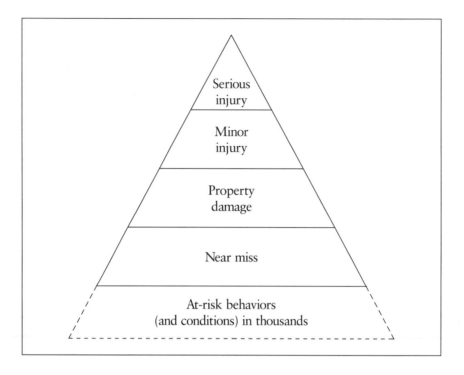

Figure 8.1. The accident ratio pyramid.

property damage accidents, and near misses.* Underlying this pyramid are thousands of unsafe acts. Heinrich estimated that 88 percent of accidents (or about five of every six) are the result of unsafe acts, which Frank Bird and the Du Pont Corporation have corroborated. These acts or behaviors, therefore, are the *common pathway* to accidents. A major failing of safety regulation in the United States and Canada has been its disregard of this axiom. *Unhappily, safety regulations emphasize unsafe conditions.*

---

*Heinrich is the originator of the accident pyramid concept. Accident ratios vary with industry.

Behaviorism—A Brief History

BBS is a product of the school of psychology known as behaviorism, which was founded in 1913 by John Watson. He defined *behaviorism* as the school of psychology that emphasizes the study of overt, measurable behavior and believes that behavior is largely the product of learning, primarily through reinforcement and punishment.

The analysis of stimulus and response is at the heart of behaviorism. Watson's central ideas were as follows:

- The measurement of overt behavior is a cornerstone of psychology.

- Human behavior is almost entirely learned.

- Any input from the environment can be a stimulus; any resulting behavior, the response.

- Because humans have few innate behavior patterns, the human mind can be considered a *tabula rasa,* or blank slate, on which the environment writes to create personality.

- Because behavior is learned, you can eliminate undesirable behavior and create desirable behavior.

These ideas remain controversial. Some—for example, the ideas that the human mind is a blank slate and that all behavior is learned—have been discredited.

Behaviorism was refined by B. F. Skinner. According to Skinner, consequences are powerful determinants of behavior. In particular, four types of consequences can be used to encourage or discourage behavior.

- Positive reinforcement

- Negative reinforcement

- Punishment

- Extinction

The first three need no explanation; extinction means withdrawing a reward in order to eliminate a negative behavior.

In the 1950s psychologists began to apply these principles in industry, schools, and the penal system, with some success in industry, but less in schools and prisons.

Limitations of Behaviorism

Some of the tenets of behaviorism remain controversial, and there are obvious limitations. For example, how does behaviorism explain the willingness of people to die for a cause? Or the selfless sacrifice of a parent for a child? Clearly, behaviorism does not apply when strong attitudes or values such as patriotism, commitment to a cause, or love are involved. It is most useful in low-voltage situations where attitudes are not particularly strong. This may explain why it has failed in settings such as schools and prisons. Behaviorism was savagely satirized in the Stanley Kubrick film *A Clockwork Orange*. Its limitations must be kept in mind in the sections that follow.

Application to Safety

In 1978, Judith Komaki, a psychologist, began applying behavior psychology to safety problems at a food-processing plant. Komaki

- Defined desired safety-related behaviors in clear simple terms
- Assessed observed behaviors against defined desirable behaviors
- Introduced the concept of "percent safe" for observed behaviors
- Provided feedback to the workers based on observed safety-related behaviors[2,3,4]

Komaki's techniques were refined by Krause, Hidley, and others, and BBS was born.[5,6]

ABC Analysis

Antecedent-behavior-consequence (ABC) analysis is a key diagnostic tool of BBS. It helps explain why employees willfully take risks and how systems can be adjusted to encourage safe behavior. ABC analysis is relatively easy to grasp but has a number of limitations. In particular, it can oversimplify human motivation and ignores individual differences. ABC analysis is a useful tool provided these caveats are kept in mind.

Consequences Are More Powerful Than Antecedents

According to the ABC model, consequences are more powerful than antecedents in determining behavior—a surprising statement. How can something that happens after the behavior influence the behavior more strongly that something that happens before it? Here are some examples.

Telephone ringing. When the telephone rings, why do people answer it? Is it because of the antecedent—the ringing? Or is it because of the consequence—someone to talk to on the other end?

Threats. A threat (antecedent) is only as strong as its potential consequences. The strong antecedent—tone of voice, body language, list of penalties—will not change behavior if there are no significant consequences. By contrast, even a mild antecedent will have a powerful effect if the consequences are certain.

Promises. A promise (antecedent) is only as powerful as the consequences it predicts.

Assessing the Strength of Consequences

Three qualities determine the strength of consequences.

- **Timing** (sooner/later): Is the consequence immediate or is it delayed?

- **Consistency** (certain/uncertain): Is the consequence likely or unlikely to occur?

- **Significance** (+/−): Is the consequence positive or negative?

The most powerful consequences are those that are soon, certain, and positive. The weakest are those that are late, uncertain, and negative.

Safety Attitudes Do Not Reliably Predict Behavior

Traditional safety management depends heavily on developing positive safety attitudes in the workforce. The underlying assumption is that those who have a positive attitude toward safety will not engage in risk-taking behavior; hence, the emphasis on safety meetings, posters, slogans, and so on. Typically, however, these efforts fall flat. Experience demonstrates, sadly, that even workers with excellent safety attitudes engage in risky behavior.

Attitude is an antecedent and, thus, is not a reliable predictor of behavior; nor is it amenable to measurement. How do you reliably measure a person's attitude toward safety? In fact, most people have positive safety attitudes. Most drunk drivers believe that it is not right to drink and drive. Most people have a positive attitude toward weight loss or giving up smoking or any number of vices—but they can't change their behavior. Why? Because a resolution is only as effective as the result (consequence) that it achieves.

The ABC Process

The ABC process works well in team brainstorming sessions.

Pick a Behavior to Analyze

Pick a high-risk behavior such as the one described at the start of the chapter. Be as specific as possible. For example, instead of writing "Failure to follow electrical switching procedures," one might write "Failure to check high-voltage indicator" and other specific behaviors. Typical behaviors could include the following:

- Failure to lock out equipment
- Not wearing protective equipment

- Standing "in the line of fire" (for example, under a suspended load)
- Poor ergonomic position (for example, excessive bending or twisting)

Write the behavior in the middle of a flip chart or on a white board.

Identify Antecedents of the Behavior

On the left side of the behavior, write *Antecedents*. Antecedents are things that come before the behavior. List possible antecedents. Some common antecedents of at-risk behavior are

- Inadequate training ("don't know any better")
- Common practice ("everyone does it")
- Protective equipment unavailable
- No time to do the job properly
- Time of day (before break, at end of shift)
- Layout or design of plant, equipment, or tools
- No other way to do the job
- Low perception of risk

Review the List of Antecedents

In the review, ensure that they are antecedents and not consequences. For instance, suppose the at-risk behavior is "not wearing hearing protection." Someone might suggest that "discomfort" is an antecedent. But discomfort is really a consequence. The actual antecedents here might be either of the following:

- Design of the hearing protection
- Condition of the ears (an infection might make wearing hearing protection uncomfortable)

Ensure that antecedents cited really do occur. For instance, "lack of training" might be listed as an antecedent, but it may not be true.

Identify Consequences

On the right side of the behavior, write *Consequences*. Ask: "What may happen to a person as a result of the at-risk behavior?" List all the things that *could* happen regardless of whether they *will* happen.

There must be positive consequences on the list. People do not do things otherwise. Positive consequences include getting something you like and avoiding something you dislike.

Not every antecedent has a consequence associated with it. However, we may get ideas on possible consequences by looking at the antecedents.

Common consequences for at-risk behavior include the following:

- Saving time

- Saving energy

- Comfort

- Injury

- Reprimand and other forms of discipline

- Avoiding hassles with coworkers or boss

- Recognition for getting work done quickly

- Feeling like part of the group

Assess the Power of Each Consequence

Determine the strength of each consequence from the three qualities below.

- **Timing:** Sooner or later

- **Consistency:** Certain or uncertain

- **Significance:** Positive or negative (+/–)

Rate each consequence on the list. Here are some possible results.

- SC+: Soon, certain, positive

- LU–: Later, uncertain, negative

- SU–: Soon, uncertain, negative

Circle the consequences that were rated SC+. These usually encourage the at-risk behavior. Traditional safety usually relies on consequences such as the potential for injury and discipline to influence the worker. But these are generally rated as LU–. *SC+ consequences generally outweigh LU– consequences.*

Here, then, is the reason that workers take risks. The utility worker in the example took shortcuts because there were SC+ consequences to doing so—including more free time for coffee and for personal business (he was an off-site worker), and peer approval. The risk of injury, by contrast, was a SU–. After all, he had been doing such work for more than a decade and had never been injured.

Consider How to Change Consequences and Antecedents

ABC analysis also suggest methods of changing behavior. Review each antecedent on the list. Can it be changed?

Now review each consequence. How might the consequences that encourage risk taking be changed? If the consequence cannot be changed, are there other consequences that can be added that might encourage safe behavior?

In the example cited, the key was to eliminate the SC+ consequences of taking shortcuts and to set up SC+ consequences for safe behavior. Punishment in the form of reprimands or increased supervision and control (both SC–) would be less effective because such punishment fails to address the root cause of the problem: The system provides SC+ consequences for risk-taking behavior. Such measures would also negatively affect morale and labor relations.

Here is an example of a completed ABC analysis.

Behavior: Not wearing hearing protection in noisy areas.

Antecedents	Consequences (S/L, C/U, +/–)
Not available in work areas	Hearing loss LC–
Area not posted	Saves time SC+
Lack of training	Discipline SU–

Won't be in area long	More comfortable SC+
Other people don't wear them	Can hear equipment SC+

What is the worker likely to do? Because there are more SC+s than SC–s, clearly, the worker will *not* wear the hearing protection. The system encourages at-risk behavior. How would one change the system to encourage safe behavior?

Behavior-Based Safety in Action

The objective of behavior-based safety is continuous, statistically significant improvement based on the measurement of operationally defined critical behaviors. ABC analysis is its key diagnostic tool. BBS requires the commitment of top management and the participation of employees. There must be a guarantee that employees will not be blamed for at-risk behaviors (that is, a no-fault system).

The main steps in the BBS process are

- Identify site-specific critical behaviors and conditions and develop a critical behaviors inventory (CBI) and a critical conditions inventory* (CCI).

- Train employees in observation and feedback skills, in how to chart percent (%) safe, and in how to deal with conflict.

- Prepare % safe charts and post them prominently in the workplace.

- Apply the PDCA cycle to solve problems. Continuously improve.

Employees must play a central role in developing the CBI and CCI, and in the observation process.

---

*Though critical conditions are not part of behavior-based safety per se, it is useful to track them at the same time as critical behaviors.

The Quality Feedback Loop

As discussed in chapter 3, a worker needs to know three things in order to produce quality.

1. The quality goal

2. The actual quality of what is being produced

3. A means of adjusting the process in the event of nonconformance with the goal

We shall see how BBS satisfies this all-important feedback loop.

Critical Behaviors and Critical Conditions Inventories

As noted earlier, five out of every six accidents are the result of an at-risk behavior. A behavior is critical to safety performance if the chance of an accident substantially increases when it is performed incorrectly. By identifying critical behaviors and controlling their antecedents and consequences, their frequency can be reduced. Similarly, critical conditions are those whose presence substantially increases the chance of an accident. One in six accidents is the result of a critical condition.

The CBI and CCI are developed by

• Reviewing accident investigation reports and identifying the behaviors and conditions that caused the accident

• Calling on the experience and knowledge of veteran workers and supervisors

• Brainstorming

The Pareto principle must be obeyed: Concentrate on the critical few behaviors and conditions. The CBI should comprise no more than about 15 behaviors. The same applies for the CCI.

The CBI and CCI correspond to the quality goal, the first part of the feedback loop.

Observation and Feedback

A check sheet should be used to document observations. The sheet should not include the name of the person being observed. Behaviors and conditions should be grouped by categories, with the most critical at the top of the section. There should be space to record comments. All observations should be done openly and with the permission of the worker. The observer should record the number of safe and at-risk behaviors and conditions. Barriers to working safely should be recorded in the comments section.

When the observation is complete, the employee should be given feedback. The observer should be positive and should emphasize behaviors performed safely. He or she should avoid judging or trying to force change on anyone. Any at-risk behaviors should be discussed in a neutral manner, along with any ideas the employee might have to prevent them. This SC+ feedback helps to tilt the mix of consequences toward safe behavior. In addition, it helps to satisfy the second part of the feedback loop, knowing the actual quality of what is being produced.

Charting % Safe

Charting % safe also satisfies the second part of the feedback loop. In addition, it harnesses the Hawthorne effect* and people's natural curiosity to know how they are doing.

Barriers to Safe Behavior

There are a number of possible barriers to safe behavior including

- Substandard facilities, including clutter that creates tripping and falling hazards and ergonomic hazards due to poorly designed jobs or workstations

---

*The Hawthorne effect war first documented in the 1920s by Elton Manyo in his pioneering work in industrial psychology at Western Electric's Hawthorne Works. Essentially, it suggests that people work better when they are being watched (in the benign sense).

- Poorly designed tools and equipment
- Disagreement on the safest way of doing the job
- An organization's culture; some cultures place a high value on risk taking
- Personal factors, in particular a person's locus of control: Does the person believe that he has control over his fate?
- Temporary impairment due to stress, illness, or drug use
- Personal choice; a small minority of people like taking risks

Such barriers must be identified and removed. This satisfies the third part of the feedback loop, the ability to adjust when the quality goal is not met.

Problem Solving

Problem solving should be done by teams comprising both employees and supervisory staff. The key is to identify the root causes of risk-taking behavior and the barriers to safe behavior.

Employees should be given training in basic problem-solving skills such as the PDCA cycle.

Over time, cooperative problem solving creates an atmosphere of trust and helps to change the culture of the organization.

Relaxed Awareness

The optimal state for safe (and productive) performance is relaxed awareness. In this state, a person is able to adapt easily to changes. It is human nature to "go on automatic pilot" when performing a task that has been done many times before. This is the *habituation* phase to which veteran workers are prone. In the absence of feedback, habituation can leave the worker "asleep at the wheel." Rookie employees have low levels of relaxed awareness—not because of habituation, but because of inexperience. Traditional safety tries to induce a state of "hyperawareness." But hyperawareness will not result in safe behavior. On the contrary, psychiatrists recognize hyperawareness as a malady.

Consider the task of driving home from work. It may be rush hour in the rain; there may be an aggressive driver tailgating you; your gas level may be low. Relaxed awareness means calmly taking in the information and adjusting as required. In this state, you perform safely and efficiently. By contrast, the driver who is hyperaware is likely to be a menace.

BBS, by clearly defining critical behaviors and conditions and by providing regular feedback to employees and the means to adjust, enhances the level of relaxed awareness among employees.

The Critical Mass Metaphor

In nuclear physics, when the number of radioactive particles in an unstable material reaches a critical mass, an explosive nuclear reaction can occur. Similarly, when the number of at-risk behaviors in a workplace reaches its critical mass, an accident occurs.* An important risk-management activity is, therefore, to determine a workplace's critical mass % safe and baseline % safe. (The closer a facility's baseline % safe to the critical mass % safe, the greater the risk.)

The ultimate goal of BBS is to continually improve the % safe level until there are zero lost-time injuries. Once this lofty goal has been reached, the goal becomes zero medical aid injuries and so on. There is likely to be a delay between an improvement in % safe and improvements in downstream variables.

Summary

Why do workers choose to take risks? Behavior-based safety helps to answer this question and to determine how systems can be adjusted to encourage safe behavior. BBS is the key to upstream measurement and

---

*Introduced by Thomas Krause.

continuous improvement in safety. Behavior-based safety requires a supportive culture.

Behavior-based techniques can also be applied to quality and environmental management. The CBI and CCI can be readily be defined according to quality or environment. The observation, feedback, charting, and problem-solving processes would be identical.

BBS is a cornerstone of the safety management system, but it is not a panacea. As noted, it is unreliable when attitudes are strong. In addition, it ignores individual differences and is impracticable when there is mistrust between the workplace parties.

Notes

1. H. W. Heinrich, *Industrial Accident Prevention* (New York: McGraw-Hill, 1931).

2. J. Komaki et al., "A Behavioral Approach to Occupational Safety: Pinpointing and Reinforcing Safety Performance in a Food Manufacturing Plant," *Journal of Applied Psychology* 63(4) (1978): 434–445.

3. J. Komaki et al., "Effect of Training and Feedback: Component Analysis of a Behavioral Safety Program," *Journal of Applied Psychology* 65(3) (1980): 261–270.

4. J. Komaki et al., "The Role of Performance Antecedents and Consequences in Work Motivation," *Journal of Applied Psychology* 67(3) (1982): 334–340.

5. T. R. Krause and J. H. Hidley, "Managing Safety Means Focusing on Behavior," *PIMA Magazine* (February 1988).

6. T. R. Krause, J. H. Hidley, and S. J. Hodson, "Behavioral Science in the Workplace: Techniques for Achieving an Injury-Free Environment," *Modern Job Safety and Health Guidelines* (New York: Prentice Hall Information Services, December 1988).

---------- CHAPTER 9 ----------

Measurement of Safety Performance

What gets measured gets done.

—Management proverb

In this chapter, the quality toolbox is applied to safety measurement. Some familiarity with SPC and other quality tools is assumed on the part of the reader. The reader who is not familiar with quality tools is strongly advised to acquire a primer. The following are excellent and may be considered as companions to the chapter.

- *SPC for Everyone* by John Burr (ASQC Quality Press)
- *The Quality Toolbox* by Nancy R. Tague (ASQC Quality Press)
- *The Memory Jogger II* (Goal/QPC)
- *Waste Chasers* (Conway Quality)

Why Measure?

Why measure safety performance? The most important reason is alluded to in the quotation above. We measure to persuade senior management to make the first big leap of commitment and to maintain that commitment. If we do not measure, safety may be lost in the shuffle. Once we have data, however, the case for safety management is persuasive on the basis of cost alone. There are also other reasons to measure.

Measurement as a Benchmark

We rely on measurement to tell us whether we are improving or not. Safety expenditures are substantial in many industries. Is the company getting value for money? How does it compare to other companies in the industry? Are there meaningful trends?

Measurement for Accountability

Measurement helps establish accountability. Organizational arrangements defined in management systems set targets against which performance in safety may be assessed. No measurement, no accountability.

Measurement as Feedback

As we have seen, quality performance requires

- A clear quality goal
- Regular feedback on the actual quality being produced
- A means of adjusting when the goal is not met

Therefore, measurement is feedback; and feedback

- Reinforces good performance
- Corrects substandard performance
- Enhances awareness

Poor performance can often be traced to a faulty feedback system.

Feedback is most useful when it provides an early warning. For example, finding out that a product is defective after it has left the plant is of little value. The external failure costs are likely to be substantial. Finding out on the loading dock is better, but the cost of taking the product apart and fixing it (internal failure costs) will still be high. Quality cost accounting shows that failure costs generally decrease as you move upstream. Upstream measurement, therefore, is the most valuable.

Current Measures of Safety Performance

Frequency Rate and Severity Rate

The most commonly used safety measures are accident frequency rates (FR) and severity rates (SR). FR and SR are the recommended measures when the number of total hours worked each month varies.

FR is defined as

$$\frac{\text{Number of incidents} \times 200{,}000}{\text{Hours worked}}$$

SR is

$$\frac{\text{Total number of days lost} \times 200{,}000}{\text{Hours worked}}$$

The term 200,000 corresponds to a workforce of 100 working 40 hours per week, 50 weeks per year. The FR indicates the percent of the workforce likely to be injured during a working year. To wit, an FR of 10 indicates that 10 percent of the workforce were injured during the year. FR and SR are usually tracked for lost-time injuries. It is also valuable to track FR for medical aid injuries and first aid injuries.

The widespread use of FR and SR allows performance to be tracked over many years. The National Safety Council's *Accident Facts,* published annually, tracks these indices across many industries over decades—a luxury few fields share. This facilitates performance tracking and benchmarking. On the negative side, FR and SR are downstream measures and have little predictive value.

Lost-Time, Medical Aid, and First Aid Injuries

When the person-hours worked per month is constant, the following accident counts may be used.

- Lost-time injuries (LTI): injuries that render the injured person unable to work for a full day or more

- Medical aid injuries: injuries that require medical attention but do not result in lost time

- First aid injuries: injuries that require first aid treatment

- Near misses: incidents that did not result in an injury but could have

These measures correspond to the accident ratio pyramid (see Figure 8.1). As you move down the pyramid, measurement becomes more predictive and hence more valuable. Lost-time and medical aid injuries are tracked by most companies. Few companies track first aid injuries and near misses.

Pareto analyses of each class of injury by cause, department, job, shifts, and so on are invaluable tools. Such analysis helps guide safety system activities such as training, group meetings, and communications.

Audits

> *Oh, goody. Another audit!*
>
> —Manager anticipating an audit

Audits are used by large companies to track safety performance. These are similar to audits conducted under quality protocols such as ISO 9000. They comprise interviews, physical conditions, tours, and document reviews. The auditors may be corporate staff or third-party consultants. Management system standards may be one of the following:

- International standards such as those of the ISO

- Internal standards developed by the corporation

- Standards developed by commercial suppliers

Such audits evaluate whether the management system satisfies the audit guidelines. On the positive side, audits are useful in the developing stages of a system in that they provide structure and help organize thinking; provide reasonably objective performance standards; and thus help establish accountability.

On the negative side, the relationship between a good audit score and superior safety performance can be tenuous. Often there is no explicit requirement for performance improvement. Another problem is the paper wall. Audits can degenerate into frustrating paper chases. Few companies have information technology systems commensurate with the needs of their management systems.

Finally, audits can be a pain in the neck. A standing joke goes like this: What are the three lies that are heard during every audit? Answer:

1. Auditor: Hi! I'm the auditor and I am here to help.

2. Auditee: Gee, we're really happy to see you.

3. Auditee: That was really helpful. Please come again.

One can hardly fault managers for taking a jaundiced view of audits. The number of required audits is burgeoning. A possible solution is the one-stop audit.

The Problem with Common Safety Measures

Focus Is Too Far Downstream

The most obvious problem with FRs, SRs, and accident counts is that they are end-of-pipe measures. They tell us little about the nature of risk. By the time an injury occurs there have probably been several thousand at-risk behaviors. To apply effective countermeasures we need to know the nature of the risk and its root causes. For example, why is the use of respirators declining in the paint shop? Why are electrical mechanics taking more chances with energized equipment? How can we adjust consequences so as to encourage safe behavior? And so on.

Misinterpretation

Common safety measures are prone to misinterpretation. Because accidents are comparatively rare, there is often insufficient data to apply SPC. In addition, applying SPC to downstream safety measures is based on two assumptions that are often untested: (1) the data are independent, and (2) the data satisfy the appropriate distribution (usually Gaussian). The following red flags list conditions under which these assumptions may be violated.[1]

Red flag 1—Normality violated.

- Small number of LTIs
- Small workforce
- Low exposure hours

Red flag 2—Independence violated.

- Reporting biases—massaging the numbers to make them look good
- Systematic changes in reporting measures
- Systematic changes in hours worked, exposure levels, weather conditions, major hirings, or layoffs
- Reported-but-not-incurred (RBNI) problems

Managers should question safety statistics in such situations.

Statistical Literacy

Unfortunately, statistical literacy is not widespread among managers. Consider the following scenario: A plant has been averaging five accidents per month over the years. The corresponding c chart, shown in Figure 9.1, indicates that the system is stable. The control limits are given by the familiar expression,

$$\bar{c} \pm 3\sqrt{\bar{c}}$$

They work out to 0 and 11.7.

Now suppose the number of LTIs in the next month doubles from five to 10. A 100 percent rise! Should the plant manager be disciplined? Should consultants be hired? Is more training needed?

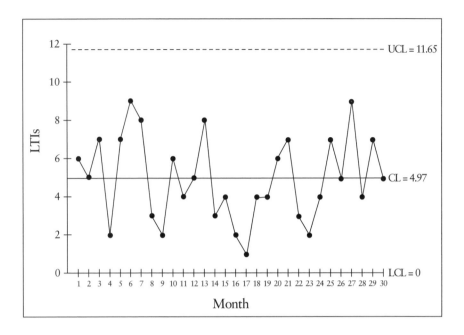

Figure 9.1. *c* chart of lost-time injuries per month.

The answers are no. This result has no statistical significance.

The absence of statistical literacy results in false feedback. False feedback leads to random responses that merely add instability to the system. These usually comprise more training (the favorite), discipline, or personnel changes.

Reliability of Data

The fixation on injuries can create pressure to underreport. For example, lost-time injuries can be reclassified as medicals or first aids. Injuries that occurred on the job are assigned to the weekend and so on. Creativity is expended manipulating the accident counts—not improving the system.

Audit results can also be unreliable. Many audit schemes allow the auditor wide discretion and do not explicitly require performance improvement. Finally, first aid and near-miss data are often unreliable because of inconsistent reporting.

The Proper Use of Downstream Measures

Frequency rates, severity rates, and other downstream safety measures can be useful if they are properly interpreted and are used together with upstream indicators, such as behavior data.

Variables Charts and Attributes Charts

Statistics essentially deals with two types of data: continuous data and discrete data (called variables and attributes data, respectively).

Variables data. Variables data are *measured* data—temperature, viscosity, pressure, and so on—and are based on *real* numbers. FR and SR data are variables data. Control charts based on variables data include $\bar{X}$–R charts, individuals charts, and X–MR charts.

Attributes data. Attributes data are qualitative—good or bad, accept or reject, and so on—and are based on integers. The number of defects per product, accidents, and at-risk behaviors are examples of attributes. Attributes control charts include c, u, p, and np charts.

Let us look at the most useful types of control charts with respect to safety measurement. Unfortunately, accident data is not amendable to the most powerful type of chart, the $\bar{X}$–R chart.

X–MR *Charts*

Monthly FR and SR results should be used when the number of exposure hours varies. The appropriate control chart is the X–MR chart. MR stands for moving range and represents the difference between consecutive results. This chart is less powerful than the $\bar{X}$–R chart but preferable to a chart of individuals.

X–MR charts are predicated on the normality of the data. Thus, before they are used, histograms should be prepared. If the data are not normal, an appropriate transformation should be made. Sometimes FR and SR data are biased toward lower numbers and the distribution is log-normal. A simple logarithmic transformation will normalize the data in such cases.

Figure 9.2 illustrates an *X–MR* chart of working days lost per month in a large facility. Figure 9.2a shows the *X* chart. There is a strong descending trend. Figure 9.2b shows the moving range of working days lost. There are several out-of-control points on both the *X* and *MR* parts of the chart.

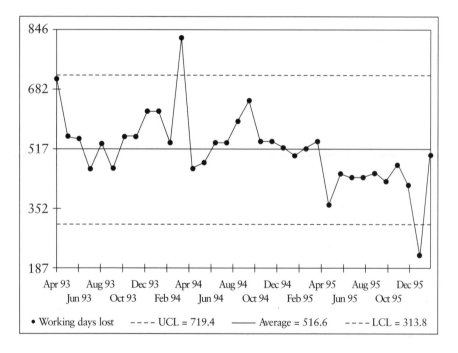

Figure 9.2a. *X* chart of working days lost.

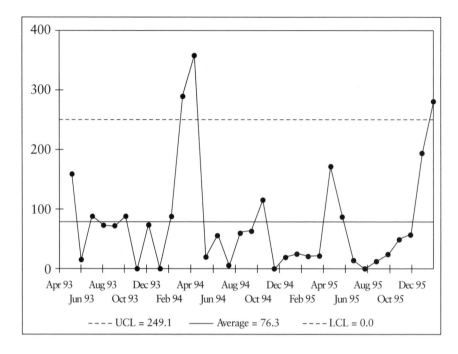

Figure 9.2b. Moving range chart of working days lost.

c *Charts*

c charts of accident counts (that is, the number of incidents of lost time, or of medical aid or first aid injuries) are appropriate when the number of exposure hours is constant. *c* charts are attribute charts based on the Poisson distribution. Poisson distributions have standard deviations equal to the square root of the mean. Once the Poisson mean reaches 10 or higher, the distribution approximates a normal distribution. Data samples with means of less than 10 are prone to misinterpretation because the control limits are misleading. Both false positives and false negatives are possible. Table 9.1 shows correct upper and lower control limits for Poisson distributions with means less than 10. Table 9.2 shows corresponding interpretation rules.[2]

Table 9.1. Correct upper and lower control limits for Poisson distributions of data with means of 1 through 9.

Mean	LCL	UCL
1	0	6
2	0	7
3	0	9
4	0	11
5	0	13
6	0	14
7	0	16
8	0	18
9	1	19

Table 9.2. Rules for interpreting runs for Poisson distributions with means of 1 through 9.

Mean	A significant low run is seven in a row equal to or below	A significant high run is seven in a row equal to or above
1	0	1
2	0	2
3	0	3
4	0	4
5	0	5
6	0	6
7	0	7
8	0	8
9	1	9

Control Chart Interpretation

A process is considered to be out of control if any of the following is true.

1. One or more points fall outside of the control limits.

2. When the control chart is divided into zones,* as shown in Figure 9.3, any of the following are true.

   a. Two points, out of three consecutive points, are on the same side of the average in zone A or beyond.

   b. Four points, out of five consecutive points, are on the same side of the average in zone B or beyond.

   c. Nine consecutive points are on one side of the average.

   d. There are six consecutive points, increasing or decreasing.

   e. There are 15 consecutive points within zone C.

The manager's job is to stabilize the process by removing special causes of variation and to gradually reduce the number of injuries by identifying and eliminating common causes of variation.

---

*Each zone represents one standard deviation from the mean.

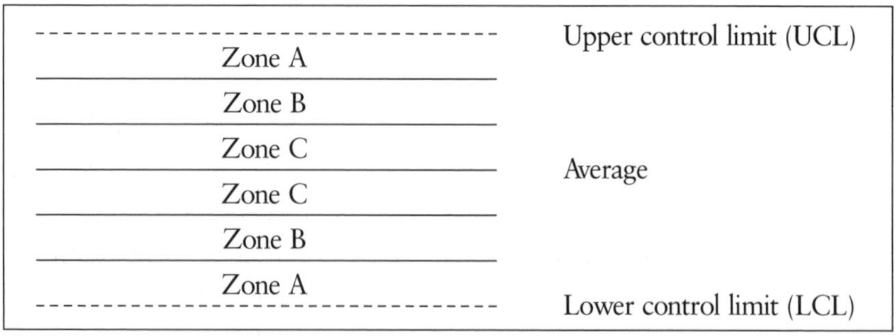

┄┄┄┄┄┄┄┄┄┄┄┄┄┄┄	Upper control limit (UCL)
Zone A	
Zone B	
Zone C	
Zone C	Average
Zone B	
Zone A	
┄┄┄┄┄┄┄┄┄┄┄┄┄┄┄	Lower control limit (LCL)

Figure 9.3. Control chart zones.

A common misconception regarding SPC is that it helps identify common causes of variation. In fact, SPC identifies special causes of variation. Deming stated that 6 percent of the variation in a process is due to special causes and 94 percent is due to common causes. SPC, then, is focused on eliminating 6 percent of the problem. Part of the disillusionment following the second wave of SPC training in the 1980s was a result of this misunderstanding. But many companies believe that eliminating that 6 percent provides the competitive edge.

Upstream Measures of Performance

Behavioral Data

Behavioral data are the key to upstream measurement in safety. p and np charts of % safe are powerful tools that can

- Establish the % safe (or % at risk) baseline for the plant.
- Estimate the critical mass of at-risk behaviors (that is, the level at which accidents are likely to occur).
- Identify and analyze trends (seasonal, shift, departmental, machine-related, and so on).
- Provide ongoing feedback to the workforce.

A large control chart of % safe (or % at risk) placed conspicuously in the work area is effective feedback. To encourage ownership, workers should plot the data themselves.

Figure 9.4 displays a p chart of % at risk data for a given facility. The control limits vary because the number of observations made per week (called the sample size, n) varies. When n varies less than 20 percent, the average sample size may be used to calculate control limits. This eliminates the onerous task of calculating control limits for each lot. Figure 9.5 displays the same data charted in this manner. There is no loss of diagnostic value.

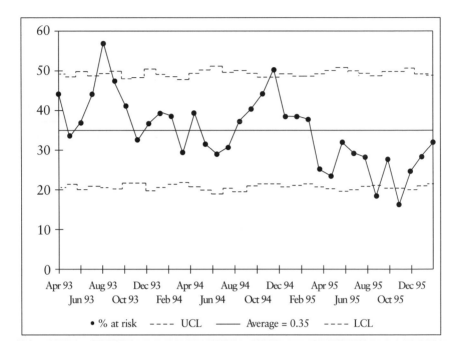

Figure 9.4. *p* chart of % at risk.

Let us analyze these charts. Clearly, there are a number of special causes at work, including the following:

- Spikes above the upper control limit in August of 1993 and November of 1994

- Six consecutive points rising between June 1995 and November 1995

- Nine consecutive points below the average between March 1995 and January 1996

- Two points below the LCL in August and November 1995

These special causes should be investigated. In general, there is a strong descending trend in the % at risk. This may be attributable

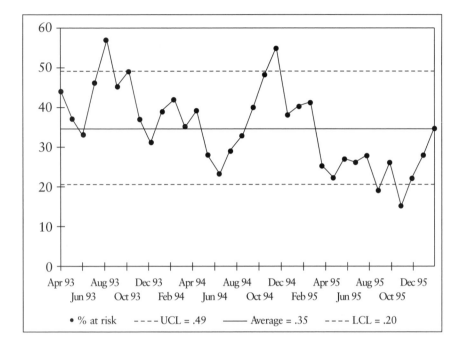

Figure 9.5. *p* chart of % at risk (using mean sample size).

to the implementation of behavior-based safety in the spring of 1993.

Figure 9.6a displays the *X* chart of accident frequency rate and Figure 9.6b the moving range. Special causes were at work.

- In February, March, and April of 1994
- Between March and December of 1995
- Between March and December of 1995

A scatter diagram relating % at risk and FR is displayed in Figure 9.7. There appears to be a positive correlation between the two indices. This was confirmed by a regression analysis.

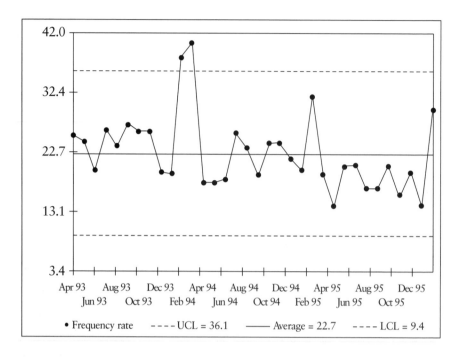

Figure 9.6a. X chart of accident frequency rate.

Future Directions

Regression Analysis of % At Risk and LTI

Here is an interesting risk analysis project. Conduct a regression analysis to determine the strength of the link between % at risk and the number of LTIs. (FR could also be used, but LTIs makes for an easier analysis.) If there proves to be a linear relationship,* determine the correlation coefficient and plot % at risk versus LTI. From the plot, determine the critical mass % at risk; that is, the value that corresponds to the onset of accidents.

---

*If the relationship is quadratic, say, and not linear, the analysis becomes more complicated but is still possible.

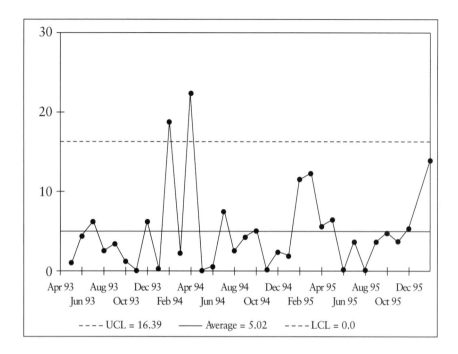

Figure 9.6b. Moving range of accident frequency rate.

Upper Specification and the z Statistic

Once the critical mass % at risk has been determined, set an upper specification for % at risk, incorporating an appropriate safety factor. For example, if the critical mass % at risk is 20 percent, the upper specification might be set at 10 percent (a safety factor of two). Now ask: Can the system satisfy the specification? Use the z value to answer this question. The z transformation formula is

$$z = (X - \mu) / \sigma$$

where

> X = value of concern (in this case the upper specification).
>
> μ = mean
>
> σ = standard deviation

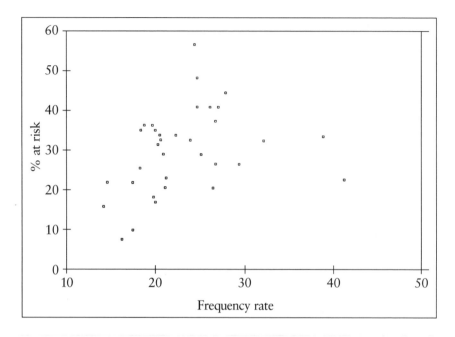

Figure 9.7. Scatter diagram % at risk vs. frequency rate.

Normality is assumed. Looking up z in a standard normal table gives the area outside the specification (displayed in Figure 9.8).

Example. Behavioral data are gathered at a large facility. The mean % at risk per month is 13 percent, σ is 6 percent, and the upper specification is 20 percent. Assume normality and a large data sample. z is equal to 1.167. The probability that a given monthly % at risk exceeds the upper specification is 0.121, as displayed in Figure 9.9. Thus, the monthly % at risk value is likely to exceed the specification about 12 percent of the time—an unacceptable level of risk. It may also be inferred that about 12 percent of critical behaviors are at-risk behaviors. Clearly, work must be done to reduce μ, the mean % at risk. In most processes, the focus is on reducing σ, but that is of less importance for behavior-based safety.

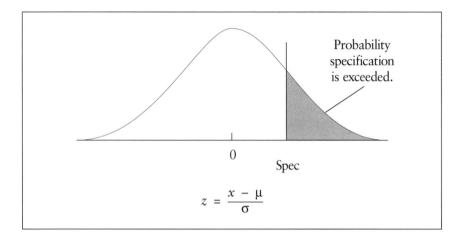

Figure 9.8. *z* value and area outside of specification.

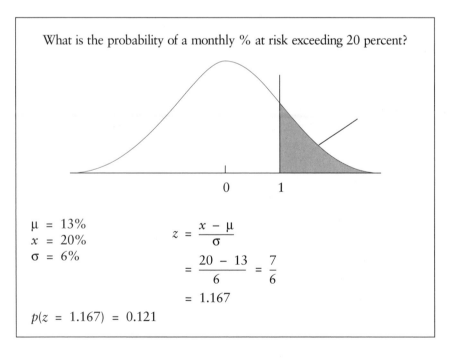

Figure 9.9. *z* value example.

Process Capability Analysis

A stable process will produce a predictable distribution of results (usually a bell-shaped curve) that can be compared to engineering specifications. A capable process is one whose distribution curve is narrower than the specifications (Figure 9.10).

The capability of a process, C_{pk}, may be defined as follows:

$$C_{pk} = (U.S. - \mu) / \sigma$$

The C_{pk} determines whether a given process can satisfy the customer's requirements and allows both producers and suppliers to quantify their risks. Table 9.3 shows the parts per million defective product corresponding to different C_{pk} values. Motorola's well-known six-sigma standard, for example, corresponds to two defective parts per billion produced. In general, a C_{pk} value of one corresponds to a process that is just meeting the specification. Values below 1 are unacceptable. Values of 2 or higher generally mean that process variation is less than the specification and is therefore acceptable.

The process capability concept can be applied to behavioral data. By determining the C_{pk} for observed behaviors in the system one can determined whether the process is capable of meeting the % at risk specification.

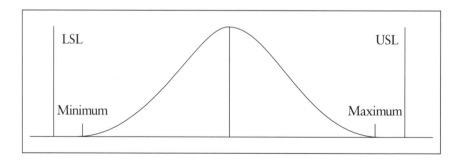

Figure 9.10. A capable process.

Table 9.3. Capability index failure rates.

C_{pk}	PPM	PPM: Parts per million of nonconformance
.33	317,500	• For a two-tailed specification
.67	45,500	• Normally distributed
1.00	2,700	• Centered on $\overline{X}$
1.10	967	• With no process shifts
1.20	318	
1.30	96	
1.40	27	
1.50	6.8	
1.60	1.6	
1.67	0.6	

Example. Is the facility described in the previous example capable of meeting the upper specification of 20 percent at risk?

$$\text{Upper specification} = 20\%$$
$$\mu = 13\%$$
$$\sigma = 6\%$$

The C_{pk} value is 0.39. One would conclude that the facility is incapable of meeting the specification. One can also infer from the z value that about 12 percent of critical behaviors (or 120,000 ppm) are at-risk behaviors.

The objective of this analysis is to point the way for future work. To the best of the author's knowledge, these concepts have never before been applied in this manner.

Summary

Accident frequency and severity rates, the most common current measures of safety performance, are standardized. This is a double-edged sword. On the positive side, benchmarking and performance tracking are simplified. On the negative side, these are end-of-the-pipe measures and assist little in the way of early warning. Other common measures include the number of incidents of lost time, medical aid, and first aid injuries. Common safety measures are prone to misinterpretation, partly due to the lack of statistical literacy among managers.

The measurement of upstream indicators is the key to continuous improvement in safety. The most useful upstream indicators are % safe (or % at risk). SPC and other quality tools should be applied to a judicious selection of both downstream and upstream measures.

The application of different types of control charts to safety measurement has been illustrated. Finally, suggestions for the direction of future work have been made.

Notes

1. Thomas R. Krause, *Employee-Driven Systems for Safety Behavior* (New York: Van Nostrand Reinhold, 1995).

2. Ibid.

CHAPTER 10

Measurement of Environmental Performance

Successful corporate environmental reporting is going to be a bit of an uphill climb for the next few years. Deal with environmental performance and communications in the same way that you would deal with any other part of your business—aim for quality and continuous improvement.

—IISD*

In chapter 9 we asked, why measure? We found that measurement

- Helps establish commitment
- Provides a benchmark
- Helps fix accountability
- Provides the feedback on which quality performance depends

In chapter 4 we pointed out that measurement is part of the quality triad, along with leadership and participation. We then introduced the PDCA cycle, which is based on an objective assessment of the current situation—that is, on measurement.

---

*International Institute for Sustainable Development and SustainAbility Ltd., *Coming Clean* (Deloitte Touche Tohmatsu International, 1993).

But in the environmental field, measurement has also taken on a strategic dimension, particularly for large multinational corporations.* Poor environmental performance can limit access to customers and to capital. A firm with a poor reputation can find its products boycotted in its own country. Foreign jurisdictions may raise various trade barriers. Indeed, nontariff barriers based on environmental performance are becoming an international trade issue.

Lending to such a firm is risky. Environmental risks are unpredictable—who could have predicted Bhopal? They are also difficult to quantify. The total cost of Bhopal is unknown and will remain so for years, until the outstanding legal actions are settled. Lenders are understandably reluctant to lend in such circumstances.

But environmental measurement is in its infancy. Surprisingly, there is no well-established way of measuring how well a firm is doing. Regulatory benchmarks tell us comparatively little. Percent compliance, number of violations, fines paid, and so on are end-of-the-pipe measures. We would like to know what is happening at each point in the system.

Corporate environmental reports are now scrutinized by various audiences including employees, investors, creditors, regulators, environmental groups, and the general public. Environmental measurement

---

*The strategic importance of environment to a firm depends on many things including (1) the nature of its environmental impacts, (2) its size and the number of countries it sells in, (3) its visibility, and (4) the importance of its corporate image. The comments made here apply mainly to large, multinational corporations. It is still possible for small and medium-size suppliers whose products are once or twice removed from the consumer (and thus "invisible") to essentially ignore environmental performance. It is enough for these firms to avoid the attention of regulators. In an era of public sector downsizing, this is not difficult to do. There is, however, a countervailing trend: the advent of ISO 14000 and other international environmental management standards. In the next century, in order to do business with large multinationals you are probably going to have to comply with such standards. This is already happening in the quality field, where ISO 9000 registration is a business requirement in some industries.

and reporting must take into account the needs of these audiences. It is not enough to focus on the practical aspects of measurement (improvement).

In this chapter we will discuss

- The audiences for environmental measurement and their needs
- The basic principles involved in choosing performance indicators
- The different kinds of indicators
- The application of the quality toolbox to these measures

Audiences for Environmental Reports

Table 10.1 lists the audiences for environmental reports and summarizes their concerns, possible environmental messages, and methods of reporting.* The complexity of the table reflects the strategic complexity of environment for a firm. Because it is probably impossible to satisfy the needs of each group, the Pareto principle must be applied. Important upstream decisions regarding the key audiences, environmental message, and reporting methods will largely determine the type and number of environmental measures chosen.

Principles of Performance Indicator Selection

Consistency with Environmental Objectives

Performance indicators must be consistent with the environmental intent, goals, and objectives set out in the environmental policy. They should also be consistent with regulatory standards. During planning, ensure that goals established are measurable and trackable. Indicators should be consistent from one report to the next so that readers can assess changes over time.

---

*I am indebted to the Canadian Institute of Chartered Accountants for this invaluable table and for the many insights provided in *Reporting on Environmental Performance* (Toronto: CICA, 1994).

Table 10.1. Environmental reporting to audiences.

Audience	Typical concerns	Environmental message	Possible primary methods of reporting
Employees	• Job security • Safety • Pay/benefits • Pride in organization	• Environmental vision and goals • Environmental achievements	• Annual report • Environmental report • Posters on bulletin boards • Video from CEO • Employee newsletter (corporate, division, plant)
Investment community	• Financial performance • Full reporting of liabilities • Prevention of future liabilities	• Risk management • Savings through improvements	• Annual report • Environmental report • Quarterly newsletter • Interviews in financial media • Surveys • Press releases/briefings
Creditors	• Assumption of environmental liabilities	• Risk management • Responsible environmental management	• Annual report • Environmental report • Interviews in financial media • Press releases/briefings
Suppliers and customers	• Product quality • Product cost • Product safety • Product liability	• Committed to delivering safe and environmentally responsible products • Available to assist with customers' safety and environmental concerns • Being supplier of choice • Goals regarding purchasing	• Advertising • Product literature/labels • Sales calls • Customer newsletter • Toll-free information line • Letters to major suppliers • Surveys

Reprinted with permission from *Reporting on Environmental Performance,* 1994, The Canadian Institute of Chartered Accountants, Toronto, Canada. Any changes to the original material are the sole responsibility of the author and have not been reviewed by or endorsed by the CICA.

Table 10.1. *(continued).*

Audience	Typical concerns	Environmental message	Possible primary methods of reporting
Community	• Pollution (health effects) • Knowledge/ understanding of organization activities • Land use ("not in my backyard")	• Pollution reduction efforts • Responsible waste management • Responsive/ concerned neighbor	• Plant tours • Neighbour newsletter/report • Speakers bureau • Visitor center • Press releases/ briefings • Focus groups/ advisory panels
Government	• Compliance with laws and regulations	• Responsible organization • Pros or cons of a given issue	• Statutory returns (public domain) • Statutory returns (confidential) • One-on-one inter-action • Formal comments • Trade association efforts
Industry associations	• Compliance with code of conduct • Public information	• Commitment to industry best prac-tices and standards	• Annual report • Environmental report • Reports to indus-try association
Activists	• Facility/community level information • Ecosystem viability • Waste elimination • Global warming • Ozone depletion • Land use	• Goals—Commit-ment to continuous improvement • Interest in working together for greater good • Performance improvement	• Annual report • Environmental report • One-on-one inter-action • Statutory returns (public domain) • See Community
Media	• See Community, Activitists, and Investment community	• Pollution reduction efforts • Success stories	• Press releases/ briefings • Plant tours/open houses • Advertising

Reprinted with permission from *Reporting on Environmental Performance*, 1994, The Canadian Institute of Chartered Accountants, Toronto, Canada. Any changes to the original material are the sole responsibility of the author and have not been reviewed by or endorsed by the CICA.

Relevance to the Audience

Indicators should provide information relevant to the audience. Audiences are likely to differ in this regard. Process data, meaningful to an operating manager, may be meaningless to a creditor. By contrast, the creditor is likely to be much more interested in data relevant to site contamination such as the condition of underground storage tanks.

Understandability by Audience

Environmental data should be translated into indicators the audience can understand. Absolute measures of energy used in terms of gigajoules, for example, are unlikely to mean much to shareholders. Instead, translate such data into everyday terms such as the number of houses that could be heated for a year with that much energy.

Selective Indicators

The number of indicators should cover the organization's most significant environmental aspects. They should be comparatively few in number so as not to overload the audience. Yet the indicators should accurately portray the costs and benefits—to the organization and the environment—of pursuing the stated goals and objectives.

Provide a Balanced View of Performance

The indicators should provide a balanced view of performance. Bad news should be reported as well as good news. The former should clearly summarize the nature of the problem and the countermeasures being taken to address it. Subsequent reports should provide updates as to progress.

The results at each point in the system—not just outputs—should be addressed. Traditional costing systems are not helpful in that they account for environmental costs as general overheads. Thus, the true cost of a given activity is often difficult to assess. The accounting profession is working to address these deficiencies.[1] As a general rule, environmental costs should be fixed at the point of origin. Secondly, the full costs of an activity should be determined using the tools of full cost accounting. (More on FCA in the next chapter).

Types of Environmental Measures

Environmental indicators may be categorized into two main groups: absolute and relative.[2]

Absolute Indicators

These are absolute measures such as total mass, total volume, total energy, and so on. Examples include total mass of solid waste generated, units of energy used, volume of water discharged, and dollars spent on improved technologies. Such indicators provide an overall picture of the resources used by a firm and its impact on the environment. They are the easiest indicators to calculate; they involve simple addition of data for the reporting period. However, they are difficult to use as industry benchmarks because the size and nature of operations vary widely. For example, suppose a firm reports that it generated 10,000 kg of solid nonhazardous waste in a given year. In this good or bad? There is no way of knowing. Suppose another firm in the industry generated 100,000 kg of nonhazardous waste. Is the second firm a worse performer than the first? Again, there is no way of knowing. It may, in fact, be a better performer per unit of production. It may be improving by leaps and bounds over time, whereas the first firm's performance is stagnant. Additional information is required: the size of each firm, performance trends, the technologies used, and so on.

Nonetheless, absolute indicators can come as a shock to firms that have never tracked them before. Polaroid executives were nonplussed to find that their firm was the leading emitter of toxic chemicals in Massachusetts. But they were dismayed when Greenpeace erected banners to this effect on highways leading to the plant.[3]

Relative Indicators

These relate absolute measures to a second measurement, usually units of production or time. Examples of relative indicators include total waste produced per unit produced, energy used per hour of operation, and so on. Such indicators are most useful in that they allow interfirm comparisons. They are likely to form the basis of industry benchmarks.

Other Indicators

Surrogate indicators. Some firms have developed indices that summarize complex scientific or engineering data in a single indicator. For example, input and output data can be weighted according to relative risk or demonstrated impact. The various inputs and outputs can then be summed into a surrogate indicator and tracked. Surrogate indicators can help assess intrafirm performance over time or between similar business units. But unless they are industry standards, which is unlikely at present, they cannot be used to compare performance between firms.

Audit results. Audit results can also be a useful measure of environmental performance. A number of firms have developed internal auditing protocols that address the EMS as well as environmental impacts. Like surrogate indicators, audit results can measure intrafirm performance over time or between similar business units. Auditing systems are also provided by independent consultants. As above, unless they are industry standards, which is unlikely, they cannot be used to compare performance between firms.

Focus of Measurement

Performance indicators can focus on inputs, the process, outputs, or impacts.

- **Inputs:** Energy, raw materials and other resources, land
- **Process:** Improved process control, improved technology, level of employee training, improved recycling

- **Outputs:** Environmental effect of products, by-products, and services

- **Impacts:** Air, water, and soil discharges; solid and liquid waste; noise; odor and dust emissions

Table 10.2 displays typical environmental performance measures found in corporate reports.

Environmental Measurement Issues

Availability of Data

Environmental indicators should be primarily based on data from mainstream data systems. These include the following:

- **Production system:** air and water emissions, groundwater and soil testing, testing of underground storage tanks, facility reports

- **Materials accounting system:** materials usage and storage, transportation of hazardous materials

- **Financial system:** operating and nonoperating capital expenditures, costs of noncompliance, research and development expenditures

Information not captured in existing systems can be designed into the EMS information system. EMS software is available that can be linked to existing business information systems.

Reliability of Data

Data availability affects the reliability of environmental indicators. Even when data are available, sampling techniques and assumptions used to form estimates can reduce the accuracy and precision of results. Appropriate margins of error should be disclosed when such disclosure could affect how the results are interpreted.

As noted in chapter 9, audit results can also be unreliable. Many audit schemes allow the auditor wide discretion. Some do not explicitly require performance improvement.

Table 10.2. Typical environmental performance measures.

Frequently used measurements and indicators found in environmental reports.		
Aspects of environmental performance	Common unit of measurement	Typical indicators
Environmental expenditures and liabilities	Dollars	Total dollars spent Total estimate site restoration costs
Compliance • Noncompliance situations	Number	Percentage of compliance Number of noncompliance situations Volume of spills
Environmental management • People trained • Environmental audits	Number	Ranking of effectiveness of implementation of programs
Technology • Investment • Waste reduction	Dollar Weight, volume	Percentage of production from nondamaging, sustainable operations
Resource usage • Material, water, other • Renewable resources • Energy	Weight, volume Area, number Gigajoules	Industry energy intensity index
Water pollutants	Weight	Total discharges Total suspended solids Biochemical oxygen demand Absorbable organic halogens
Air pollutants	Weight	Total emissions Particulates
Waste	Weight, volume	Total volume of waste Toxicity rating
	Area Number	Rate of illness Concentration of substance in living things

Reprinted with permission from *Reporting on Environmental Performance*, 1994, The Canadian Institute of Chartered Accountants, Toronto, Canada. Any changes to the original material are the sole responsibility of the author and have not been reviewed by or endorsed by the CICA.

Absence of Common Definitions

Finally, a major limitation of environmental indicators is the absence of common and consistent definitions. This all-pervasive problem exists in regulatory, financial, and industry reporting practices.

Regulatory. One would expect regulatory indicators to be straightforward. But compliance can be misleading if, for example, it is based on a jurisdiction that allows high emission levels. Secondly, percent compliance can be based of the number of readings required by regulation or the total number taken by the organization.

Financial. Financial indicators can also be misleading. GAAP* does not yet have consistent rules relating to environmental costs and liabilities. A 1992 survey by Price Waterhouse found that corporate accounting practice differed significantly in areas such as whether certain types of environmental costs should be capitalized or charged immediately to operations, when to recognize liabilities related to the clean-up of contaminated sites, and so on.

Industry standards. It is not surprising that industries are also having difficulty coming up with common performance measures. In general, common industry benchmarks do not exist. The reader of a corporate environmental report does not know if the organization is an environmental leader or a laggard. Common measures and benchmarks such as best practice standards are essential if environmental reports are to be credible.

Applying the Quality Approach

> *Current status, desired status, countermeasures.*
>
> —The PDCA mantra

The quality approach may be applied to all of the environmental measures discussed. Environmental parameters are just one more system output. The outputs of the paint shop in a large automotive plant include the number of units painted, the number of defects per car,

---

*GAAP stands for generally accepted accounting principles, which guide the accounting profession.

and so on—as well as the solids content of waste water and the quantity of volatile organic chemicals (VOC) going out the stack.

The PDCA cycle and its mantra—current status, desired status, and countermeasures—should be the drumbeat informing all environmental improvement efforts. Use the quality toolbox and the problem-solving process described in chapter 4. Here is a brief description of the process at work.

Use flowcharts to understand the key processes. In a brainstorming session, use Ishikawa diagrams to identify the root causes of the problem and possible parameters to measure. Design user-friendly check sheets to collect the data. Plot the data on a run chart to identify trends. Was the measured parameter important? If not, go back and pick another.

Use Pareto diagrams to identify the critical few causes. Use scatter diagrams to link key process variables. Apply SPC and continually improve the process.

Use SPC to maximize environmental control processes such as wastewater treatment and scrubbers. SPC can also be applied to impacts such as spills and air and water emissions. Spills are nonconformities and may be studied using c charts in much the same manner as accident counts. Air and water emissions are variables data and may be charted using X–MR charts. The comparative scarcity, at present, of input, output, and impacts data limits the application of SPC. But this is a short-term problem.

Figure 10.1 displays a c chart of spills per month. The interpretation rules outlined in the previous chapter in Tables 9.1 and 9.2 apply here. There appears to be a strong downward trend. Figure 10.2 displays a scatter diagram of monthly spills and monthly action requests by employees. There appears to be a negative correlation between them, which suggests the program has been successful. In other words, the more action requests per month, the fewer the spills. (This was confirmed by regression analysis.)

Figure 10.3 displays an X–MR chart of energy costs per unit of production. There are out-of-control points in the second quarter of 1993

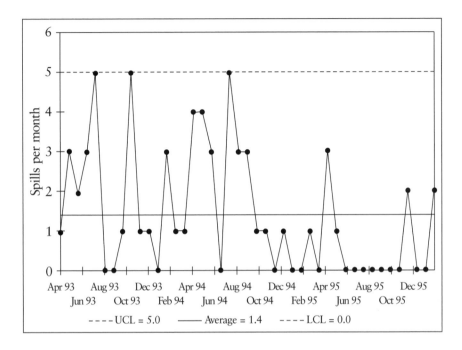

Figure 10.1. *c* chart of spills per month, ABC Oil and Gas Inc., 1993 to 1996.

and the third quarter of 1994. Then there is a strong downward trend. The important questions are: What caused the out-of-control points? Why has there been a downward trend? What are the critical few components of energy cost? What are the root causes of high-energy expenditures? What are the important parameters to track? And so on.

We are barely scratching the surface here. The potential of the quality approach applied to environment is tremendous. A core group of leading companies is beginning to apply these concepts.* But they are the exception.

---

*The Global Environmental Management Initiative (GEMI), based in Washington D.C., comprises a good cross section of these firms.

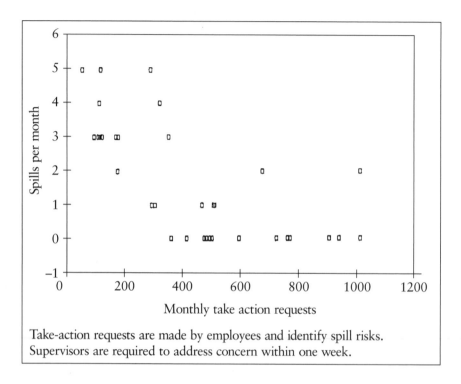

Take-action requests are made by employees and identify spill risks. Supervisors are required to address concern within one week.

Figure 10.2. Scatter diagram: Monthly spills versus take-action requests, ABC Oil and Gas Inc., 1993 to 1996.

Summary

Environmental measurement helps establish commitment, provides a benchmark and feedback, and helps fix accountability. But it is also important strategically. Poor environmental performance can limit a firm's access to markets and capital. It can affect share prices and even long-term viability.

Environmental measurement is in its infancy, and there is no well-established way of measuring how well a firm is doing. This chapter identified the many and varied audiences for environmental measurement and their needs. Principles involved in choosing indicators as well as the different kinds of indicators were also discussed.

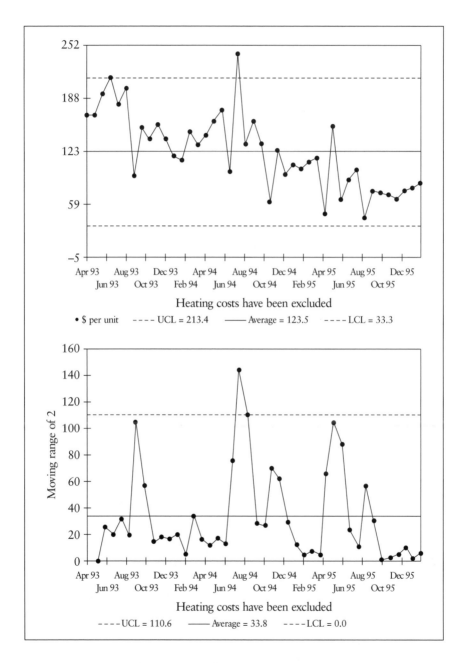

Figure 10.3. *X–MR* chart. Energy ($) per unit of production ABC Oil and Gas Inc., 1993 to 1996.

Rudimentary applications of the quality approach to environmental measurement were outlined. Environmental control has generally meant sticking something on the end of the pipe, and the process upstream of the pipe has not been a focus. But this is where the quality approach can do the most good. Its potential is limited only by the creativity and energy of the team. A group of leading companies is beginning to apply these principles.

Notes

1. Canadian Institute of Chartered Accountants, *Environmental Costs and Liabilities: Accounting and Financial Reporting Issues* (Toronto: CICA, 1994).

2. Canadian Institute of Chartered Accountants, *Reporting on Environmental Performance* (Toronto: CICA, 1994).

3. Ibid.

CHAPTER 11

The Total Safety and Environmental Management System

The devil is in the details.

—Proverb

In chapter 7, we defined total safety and environmental management (TSEM) and outlined its goals and methods. The TSEM approach allows the latent synergy between quality, safety, and environment to develop. In this chapter, we will examine the TSEM system.

Let us review some basic definitions. A *safety management system* (SMS) is an orderly set of components that serves to accomplish one or more safety goals of the organization. A corresponding definition applies for the environmental management system (EMS). The TSEM system comprises the SMS and the EMS (and their overlap) and is informed by the quality approach.

Figure 11.1 displays the generic system structure, introduced in chapter 6, that serves as a blueprint for the SMS and EMS. Figure 11.2 displays a complementary model that illustrates the relationship between the policy and core and advanced* programs. Core programs

---

*I use the term *advanced program* for simplicity. It denotes (1) programs that support core programs, (2) programs that go beyond the main goals of the system, and (3) system maintenance and improvement programs.

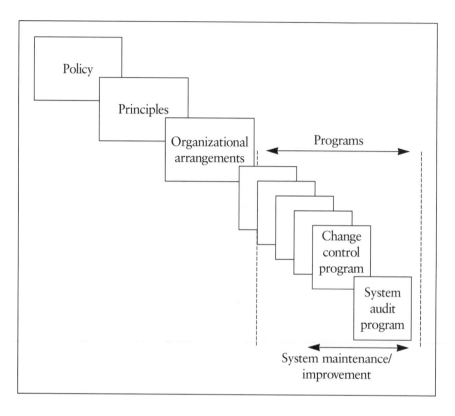

Figure 11.1. Suggested system structure.

are linked to the policy by solid lines and advanced programs by dotted lines, reflecting the relative strength of the relationship.

Figure 11.3 displays the links between the EMS and the SMS. As discussed in chapter 7, this overlapping area comprises common and crossover programs. You can also imagine a dotted line linking the environmental and safety policy. Rare is the safety policy that does not address environment and vice versa.

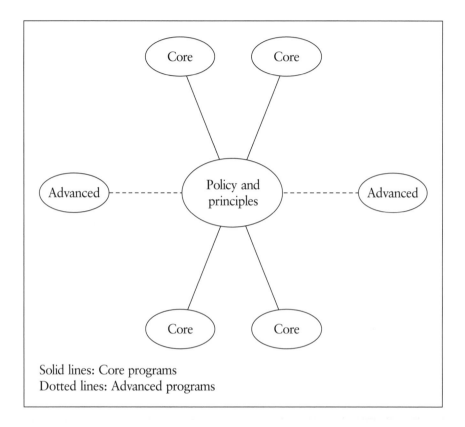

Solid lines: Core programs
Dotted lines: Advanced programs

Figure 11.2. A related system model.

The Safety Management System

The loss causation model described in Figure 6.2 should be kept in mind in the following sections. Figure 11.4 displays a generic program mix for a safety system.*

---

*A caveat is in order. This book is not a primer in safety management. Some familiarity with the field is assumed. The work of Frank Bird, Dan Petersen, and the National Safety Council is a good introduction for readers who would like to know more.

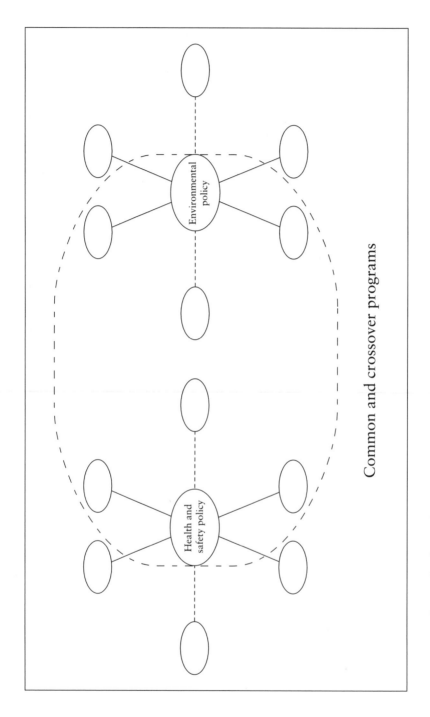

Common and crossover programs

Figure 11.3. The total safety and environmental management system.

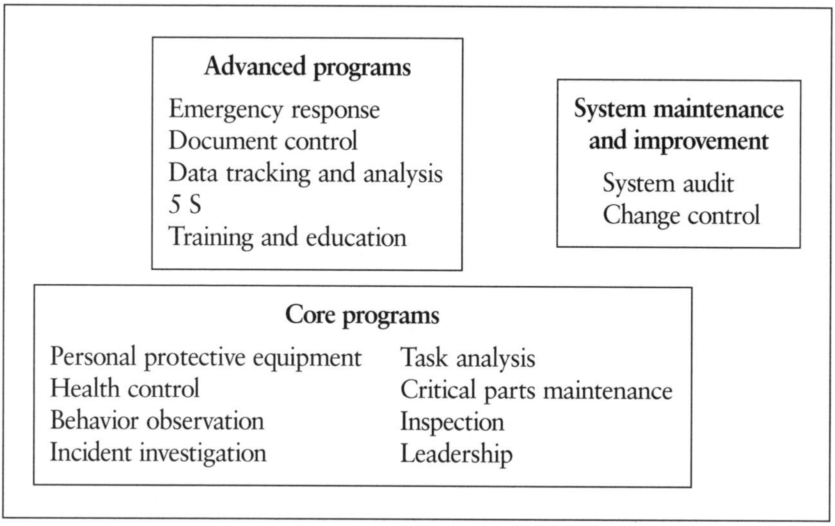

Advanced programs Emergency response Document control Data tracking and analysis 5 S Training and education	**System maintenance and improvement** System audit Change control

Core programs

Personal protective equipment	Task analysis
Health control	Critical parts maintenance
Behavior observation	Inspection
Incident investigation	Leadership

Figure 11.4. Safety maintenance system generic program menu.

Core Programs

> *Safe work is the door to all work. Let us pass through this door every day.*
>
> —Eiji Toyoda

Leadership. Leadership is the epitome of upstream control. Leadership for safety comprises the following:

1. A clear vision, simply described, widely communicated, and shared

2. A clear concise plan to achieve the vision

3. Visible and active support for the safety system

4. Management accountability

The leadership program must translate these general principles into commitment, direction, and action.

Point 1 entails a thoughtfully worded safety policy and supporting principles. These should be communicated through the following channels: wall plaques, company meetings and events, business cards, corporate reports, training sessions, company newsletters and other communications, policy manuals, and so on. A senior manager should be designated as safety coordinator.

Point 2 entails an SMS manual comprising concise descriptions of the structure and program content of the system. For each program there should be a simple flowchart, responsibility matrix, and one or two performance measures. Check sheets, report templates, historical trends, and other supporting documentation should be put in the appendix.

Point 3 requires the setting of performance standards for managers for activities such as visibility and loss control tours, incident investigations, group meetings, task force and special projects work, and so on.

Point 4 requires that managers be held accountable for their performance. This is a common breakdown point, because, in most firms, safety (and environment) are weaker corporate values than production. Managers who are productive but not safe continue to be rewarded. The message is clear to all: *Safety is not important here.* Safety efforts are crippled and cynicism grows. The safety policy becomes a source, not only of ridicule, but also of bitterness because of the gap between its lofty goals and reality. The desire for workplace safety has often been a root cause of unionization.

There is a tragic misunderstanding here. For most of this century it has been believed that safety and productivity are mutually exclusive; however, efficient production *is* safe production. Dr. Toyoda recognized this obvious but strangely invisible truth in his quotation. Perhaps it is invisible because it is so obvious.

Jim Stewart, a driving force behind Du Pont Canada's spectacular safety performance, argues that safety and profitability are the concurrent results of the same management principles and practices. A world-class company is one that achieves "balanced excellence" across a range of values including profitability, safety, and environment.[1]

Let me go one step further and suggest that a company's safety performance is an auger of its potential profitability. A company's safety record reflects the strength of its leadership and management systems, as well as its ability to manage risk, and it is a good barometer of employee relations.

Task analysis. Task analysis is the systematic analysis of critical tasks so as to identify and eliminate sources of risk and waste. Task analysis is the door to standardized work, a feature of all advanced production systems.

A critical task is one that, if done improperly, is likely to lead to injury to people, property damage, or process loss. Task analysis breaks a given task down to its components and addresses the safety, environment, and quality aspects of each. Its results will inform most other programs in the system. Its output is a series of standard operating procedures (SOP), also known as standard work charts (SWC).

Behavior observation. This program is described in detail in chapter 8. About five of every six accidents are caused by at-risk behaviors.* I will add here only that the critical behavior and critical conditions inventories should be linked with the standard operating procedures that are the output of task analysis.

Inspections. Planned inspections entail the systematic inspection of the workplace to uncover substandard work conditions. The TSEM system places less emphasis on inspections than on traditional safety: Only one in six accidents is caused by unsafe conditions. Planned inspections require much upstream work to be effective. System designers must ask: Is there any value in inspecting this area, this frequently? User-friendly check sheets are a key tool. Data generated

---

*The is one of Heinrich's original axioms industrial of safety. It has been corroborated by the work of Frank Bird and the Du Pont Corporation.

should be tracked and analyzed using quality tools. Otherwise, inspections can degenerate into mechanical walk-throughs.

Incident investigation. The purpose of incident investigation is to uncover weaknesses in the system. An incident is defined by Frank Bird as an event that has resulted in, or could have resulted in, harm to people, process loss, damage to property, or damage to the environment.

Near misses should be investigated as diligently as injuries, particularly if high-energy sources are involved. Suppose that the wire rope on an overhead crane breaks while transporting a chemical drum. If the drum falls on an employee, the employee will be seriously injured. If it falls on a piece of equipment, there will be property damage. If the drum falls in an open area and ruptures, there will be an environmental spill. Finally, if the drum falls into an open area and does not rupture, it is a near miss.

The outcomes of this incident are entirely by chance. The outcomes are less important than what the incident tells us about the system. The incident described could indicate weaknesses in the following areas.

- Critical parts maintenance
- Purchasing standards
- Inspections

The purpose of the investigation is to follow the thread back to the system weakness.

Critical parts maintenance. Critical parts maintenance (Frank Bird's term) is the systematic identification and maintenance of parts whose failure is likely to result in a major loss. Critical parts for a metal stamping machine would include the punch press brake, clutch, and guard. Critical parts on an overhead crane would include crane hooks, wire rope, and so on.

Once critical parts have been identified, maintenance requirements can be determined and assigned. Critical parts maintenance should be integrated with existing preventive maintenance activities. Software is available that greatly simplifies the planning and tracking of such activities.

Personal protective equipment. Many operations require the use of personal protective equipment (PPE). PPE is the last line of defense against contact with an energy source. Upstream control in the form of work design is not always possible. For example, how can one design out the hazards for an electrical mechanic working with live equipment at the top of a 30-foot pole? Barring major technological changes in electrical distribution, such work will require at least rubber barriers to cover live equipment, a fall-arrest system, and rubber gloves.

PPE programs identify the PPE needs of each occupation and task, and ensure that employees are properly equipped and trained in the use of that equipment.

PPE programs depend on the output of core programs such as task analysis. The nonuse of PPE is a common at-risk behavior and amenable to the behavior approach described in chapter 8.

An interesting cultural note: Some Japanese managers feel that PPE is overemphasized in North America. Protection, they point out, should be provided by the system.

Health control. Most accidental injuries occur at the moment of contact with an energy source. A trip or a fall, for example, results in a simultaneous contusion. An occupational disease, by contrast, is an injury that develops over time. Occupational disease accounts for a comparatively small percentage of workers' compensation cases but receives the lion's share of media attention. Occupational cancers are often front-page news; a worker injured in a fall is rarely newsworthy.

The rate of occupational disease continues to decline with the advent of improved engineering controls and the dematerialization of industry. It remains a major concern in the heavier industries such as manufacturing, mining, and construction.

The health control program is one of the most complex in the SMS. It covers the recognition, evaluation, and control (REC) of hazards, as well as the provision of first aid, primary health care, and more advanced medical services.

The REC triad is the basis of the industrial hygiene profession. Health control comprises subprograms to address the following:

- Chemical hazards including dust, mists, fumes, and vapors

- Physical hazards including noise, vibration, radiation, and thermal stress

- Biological hazards

- Psychological stress hazards

The last mentioned is the least understood. This should not preclude an assessment of the current condition in a facility and the development of countermeasures.

The delivery of health care services usually requires on-site medical personnel and a health center. Health center staff can play an important role in the prevention of ergonomic injuries by developing job-hardening schedules and tracking the progress of new employees. This is especially important in manufacturing, where the demands of lean production have shortened the time available for a new employee to learn and physically adapt to a job.

Ergonomics

> *Manifold is the harvest of diseases reaped by craftsmen . . . As the cause I assign certain violent and irregular motions and unnatural postures . . . by which . . . the natural structure of the living machine is so impaired that serious diseases gradually develop.*
>
> —Bernardino Ramazzini, 1713

The last core SMS program we will discuss is also the most important in many industries. Ergonomics is the science of work. Ergonomists seek to understand the person–machine interface by answering questions such as the following:

- What sort of physical burden is this job placing on the employee?
- How can this work area be improved to reduce that burden?
- What proportion of the population can work under this burden without injury?
- What are the risk factors that contribute to the ergonomic burden?
- What are the environmental factors (heat/cold stress, lighting, air quality, noise, and vibration) that might affect the performance of a person doing this work?

A mantra of ergonomics is "fit the work to the worker" rather than the other way around. Hence, the introduction in advanced companies of adjustable chairs, benches, jigs, controls, and operating panels.

Importance of ergonomics. Ergonomic injuries comprise more than half of all work injuries and are a root cause of the workers' compensation crisis described in chapter 4. More ergonomic injuries may be on the horizon. In the safety literature, there is concern that we are on the verge of an epidemic in North America of carpal tunnel syndrome, a repetitive strain disorder of the wrist and hands.

Ergonomic risk factors include

- Repetition (largely based on the process cycle time)
- Force required
- Posture
- Tools used
- Environmental conditions

The key to a sound ergonomics program is a method of integrating these factors into a meaningful burden scoring system. Unfortunately, most ergonomic measurement systems are not user-friendly.

Once a scoring system has been established, jobs can be rated according to burden and meaningful control strategies devised. Upper and lower body (back) burden should be differentiated. Each job should be color-coded according to upper and lower body burden.

Control strategies include burden reduction (based on reducing risk factors) and periodic job rotation (from high-burden to lower-burden jobs). Adjustable workstations are generally a good investment because they allow employees to tailor the work area to their physical needs.

The U.S. military has long understood the links between ergonomics and safety and has done much pioneering ergonomic research. The links between ergonomics and quality—intuitively obvious—have been empirically demonstrated.[2]

Advanced Programs

Advanced programs support core programs, enhance the performance of the system, or go beyond the core goals of the system.

The last category includes health promotion, off-the-job safety, and employee assistance programs. Although they are interesting and often effective, we will not be able to discuss them here.

Training and education. Deming's sixth point deals with training; his thirteenth with education. He differentiated between the two. The former addresses "how;" the latter, "why" as well as "how."

The mix of training and education will vary with the position. In general, managers will get more of the "why" than wage-roll employees. As we have seen, all managers must be statistically literate. In addition, all operating managers should receive education in accident theory and health and safety law. The training of wage-roll personnel will vary with their occupation. Trade qualifications standards will partly determine curricula.

A systematic approach should be used to determine training and education needs. The output of this process should be a matrix relating training needs and positions. Once the training matrix has been developed, you should (1) prioritize needs; (2) determine the appropriate teaching method; and (3) set SMART goals against which to assess training effectiveness. Refresher training should be provided to ensure that critical skills are not forgotten.

A common problem is that training can become an end in itself; it is of little value if not applied. Employees have the opportunity to apply what they have learned. A corresponding goal should be formulated and training sessions spaced accordingly.

5 S's. The 5 S's are

- *Seiri*—organization
- *Seiton*—neatness
- *Seiso*—cleaning
- *Seiketsu*—standardization
- *Shitsuke*—discipline

The 5 S's, a sophisticated housekeeping program, have been called the keys to a total quality environment. They are also the key to a total safety environment. The interested reader is referred to Osada's fine book.[3]

Data tracking and analysis. Tracking and analysis of both upstream and downstream data using the quality toolbox is a core TSEM activity. Management decisions must be based on data. Tracking and analysis of safety data were discussed in chapter 9.

Emergency response. The emergency response program (ERP) is a postcontact program designed to contain or eliminate losses once an emergency situation has begun. It usually addresses

- Fire protection and evacuation
- Spill response
- Natural disasters
- Security (bomb threat, civil disruptions or riots, criminal incidents)

The process steps are

- Identify possible emergency scenarios.
- Assess each scenario in terms of the probability of occurrence and severity of the outcome.*
- Develop responses to priority scenarios.
- Designate and train an emergency response team.
- Work out the logistics (location of control center, responsibility matrix, coordination with local authorities).

Risk assessment is at the heart of the ERP. Without it, energy is likely to be dissipated in chasing chimeras. ERPs have gained importance in the face of high-profile industrial disasters that have ravaged balance sheets, as well as people and the environment.

A common failing of ERPs is an overly complicated manual. A thick ERP manual is virtually useless in an emergency. The ERP manual should be short, simple, and easy to read. It should be hung on the wall in prominent areas and given to each employee. A storyboard format explaining what to do under each scenario works well. Less is more.

System audit

> *Say what you do. Do what you say, and show me.*
>
> —Auditors' saying

The SMS should be regularly audited to assess the adequacy of programs, the adequacy of program standards, and the degree of compliance with program standards.

System audits usually comprise interviews, documentation review, and walk-through surveys. Interviews are based on a series of questions developed by the audit team. Both managers and wage-roll

---

*Risk is the product of frequency and severity.

employees should be interviewed. Usually, there is a corresponding scoring system.

Document reviews can be frustrating for line managers. The auditor must be firm in the face of claims that "Of course we do it. It's just not documented." Remember the auditors' saying.

Walk-through surveys can be the most useful part of the process. Auditors should be prepared to roll up their sleeves and "go see," to use Toyota's popular term. It is amazing what a skillful auditor can uncover simply by being there.

Change control. Everything changes: processes, equipment, materials, buildings, and employees. The SMS will lose touch unless it is regularly updated. The components of change control include

- A procedure to notify the safety team of intended major changes

- An assessment of the safety impact of the change

- A requirement that the safety team leader signs off

- A corresponding change to the safety management system

The novelty principle. The novelty principle states that novelty equals risk. Novelty is a root cause of most serious accidents. In the context of SPC, novelty is a special cause, a major source of instability in the system. The change control program recognizes the novelty principle.

Documentation

> *Less is more.*
>
> —Robert Browning

Quality professionals involved in ISO 9000 implementation have a sophisticated understanding of documentation. ISO 9000 requires four levels of documentation to provide evidence of a system.

In chapter 6, I expressed my concerns about the paper wall. To repeat, less is more. The safety system manual should be no more than 30 or 35 double-spaced pages comprising concise descriptions of system structure and program content. The following sections would suffice.

- Safety policy and principles
- Organizational arrangements
- Program mix
- Appendix, containing forms, check sheets, and other relevant documentation

Each program should be described by a simple flowchart and responsibility matrix, and one or two performance measures. Historical trends are useful if available. Each operating department may wish to tailor the system manual to its own needs.

Legal dimension. There is a legal dimension to safety and environmental documentation. The following questions should be asked to determine the corresponding documentation required.

- What are you expressly required to do by law?
- What are you implicitly required to do by law?

The latter relates due diligence, a legal concept based on the following question: Did the person take all precautions reasonable in the circumstances? If the answer is yes, then the person is said to have shown due diligence. Documentation should ensure that the organization and individuals with duties can demonstrate due diligence under the law.

Summary

We have described the core and advanced programs in a generic SMS. Other programs can be developed in accord with the organization's needs. The challenge is to integrate them into a user-friendly system.

The Environmental Management System

An environmental management system (EMS) is an orderly set of components that serves to accomplish one or more environmental goals of the organization. Figure 11.5 displays a generic program menu for an EMS. The degree of overlap with the SMS (see Figure 11.4) is striking. Table 11.1 displays common programs and programs that have crossover potential, many of which were described in chapter 7.

Core Programs

Leadership. The points made in the previous section apply equally here. Leadership for environment requires

1. A clear vision, simply described, widely communicated, and shared

2. A clear, concise plan to achieve the vision

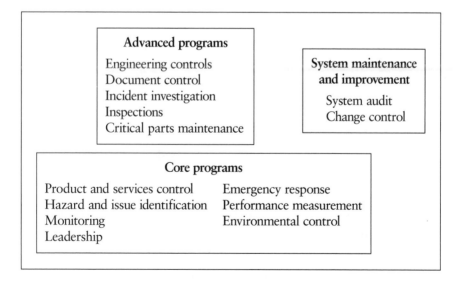

Figure 11.5. Environmental management system generic program menu.

Table 11.1. Common and crossover programs for safety and environment.

Common programs
Leadership
Inspection
Document control
Investigation
Training and education
Emergency response
Change control

Crossover programs
Task analysis
Critical parts maintenance
Behavior observation

3. Visible and active support for the environmental system

4. Management accountability

Point 1 entails an adroitly worded environmental policy and supporting principles. The policy must not promise more than can be delivered. It should be written after the firm has identified its major environmental aspects and impacts.

- Environmental aspects: the things that *can* go wrong or issues that an important audience is concerned about

- Environmental impacts: those aspects that are actually affecting the environment

For example, the environmental aspects for a mining company might include solid and liquid waste, energy and raw material use, discharges to air, water, and soil, and land use. But all of these aspects may not be generating environmental impacts. The hazard and issues identification program described below addresses environmental aspects and impacts.

Point 2 requires an EMS manual that clearly describes the structure and program content of the system. Each program should be described by a simple flowchart and responsibility matrix, and one or two performance measures. Supporting documentation should be put in the appendix.

Point 3 requires the setting of performance standards for managers for leadership activities such as visibility and inspection tours, incident investigations, group meetings, task force and special projects work, and so on.

Point 4 requires that managers be held accountable for their environmental performance. Because of absence of performance benchmarks and accounting practices that fail to fix environmental costs at the source, this is still comparatively rare. With the onset of ISO 14000 and the maturation of environmental cost accounting, accountability for environment is likely to grow.

Environmental performance does not affect workers as directly as safety, but workers want to work for responsible firms. Most workers will go the extra mile for environment if given the opportunity to do so.

Environmental control. The environmental control program addresses the company's environmental impacts. These may include the following:

- Emissions to atmosphere
- Discharges to water
- Soil and groundwater contamination
- Resource consumption
- Waste production and handling
- Land and wildlife management

Control activities include establishing countermeasures and performance targets for the major impacts. As always, heed the Pareto principle. Control the critical few impacts first. Only then should you address the trivial many.

Monitoring. Assessing environmental performance requires data on the company's impacts. The purpose of the monitoring program is to provide such data through planned sampling and monitoring at and around the firm's sites. Although the technology of environmental monitoring is beyond the scope of this book, the quality toolbox including metrology (the science of measurement) and SPC should be applied.

Performance measurement. The purpose of the performance measurement program is to analyze, interpret, and report the data generated by the monitoring program to the relevant audiences. Compliance assessment is a related activity. Chapter 10 addresses this program.

Hazard and issues identification. The hazard identification program should provide a systematic way of identifying the major environmental aspects and impacts of a firm's activities. It plays a central role during policy formulation.

Emergency response. The emergency response program is entirely congruent with that described in the SMS. One integrated ERP should be established.

Product and services control. The objective of the product and services control program is to define the environmental impacts of a firm's products and services from cradle to grave, as it were. Product and service control should address the big picture, including the impact of the firm's suppliers and the ultimate disposal of its products. Product and services control is a sophisticated program informed by the cutting edge of environmental management. Life cycle analysis, a promising new technique still in its infancy, is likely to become its core tool.

Advanced Programs

Many advanced environmental programs are congruent with the safety management programs described earlier and should be integrated. These include critical parts maintenance, inspection, incident investigation, document control, and training and education. Let us briefly describe the most important advanced EMS program.

Engineering controls. The engineering controls program has two components.

- Controls at existing facilities such as scrubbers and treatment ponds
- Control over the design of new products and processes (design for environment is a promising new tool in this regard)

The objectives of this program are to ensure that existing controls are effective and that new product and process engineering addresses environmental impacts.

Summary

There is broad overlap between the EMS and the SMS. Common programs include leadership, emergency response, inspection, incident investigation, document control, and training and education. These can effectively be integrated.

The system maintenance programs, system audit, and change control are also common and should be integrated.

Design for environment (and life cycle analysis) are core tools of a powerful new systems view of industry–environment interactions called industrial ecology. This new paradigm should inform the existing environmental management system and future improvement efforts.

Industrial Ecology

> *No man is an island.*
>
> —John Donne

No company is an island, either. Each firm is linked to thousands of people, transactions, and activities, and to their environmental impacts.

A large firm may manufacture thousands of individual products for ever-changing customers located around world. Customers will use the product according to their needs and whims. When finally disposed of, the product may end up in almost anywhere—in any country, in a modern landfill, by the side of the road, or in a lake that supplies drinking water.

Industry is skilled at identifying and satisfying customer needs. It is less adept, however, at determining the long-term consequences of how it satisfies those needs. As discussed in chapter 5, since the Age of Reason there has been a schism between humanity and the natural world. Industry—like the rest of us—has perceived itself to be separate from the web of creation.

This has created many unexpected problems. Many great industrial innovations of the past are today's environmental problems. Think of nitrogen and phosphorus fertilizers, chlorinated pesticides, tetraethyl lead, and chlorofluorocarbon refrigerants. Each solved a great problem in its day. Some saved lives. Chlorofluorocarbons, for example, replaced highly flammable refrigerants that had caused countless fires and explosions.

But there was no attempt to relate industrial processes to possible environmental consequences. Such efforts do not ensure that an activity will be environmentally friendly, but they can help avoid the worst and lay the groundwork for continual improvement.

Industrial ecology is

> *The means by which humanity can deliberately and rationally approach and maintain a desirable carrying capacity, given continued economic, cultural and technological evolution. The concept requires that an industrial system be viewed not in isolation from its surrounding systems, but in concert with them. It is a systems view in which one seeks to optimize the total materials cycle from virgin material, to finished material,*

> *to component, to product, to obsolete product, and to*
> *ultimate disposal. Factors to be optimized include*
> *resources, energy and capital.*[4]

The adjectives in this definition—deliberate, rational, desirable—differentiate industrial ecology from the more traditional forms of industry–environment interactions. These have tended to be unplanned, unpredictable, costly, and disruptive.

Principles of Industrial Ecology

Here are some of the principles of industrial ecology.

- Every molecule that enters a manufacturing process should leave the process as part of a marketable product.

- Every joule of energy used in manufacture should produce a desired material transformation.

- Industries should make minimum use of materials and energy in products, processes, and services.

- Industries should choose abundant, nontoxic materials when designing products.

- Industries should obtain needed materials through recycling streams (theirs and those of others) rather than through raw materials extraction.

- Every process and product should be designed to preserve the embedded utility of the materials used. An effective way of doing this is by designing modular equipment and by remanufacturing.

- Every product should be designed so that it can be used to create other products at the end of its life.

- Every industrial landholding or facility should be developed, constructed, or modified with the goal of maintaining or improving local habitat and species diversity and of minimizing impacts on local resources.

- Partnerships should be developed with materials suppliers, customers, and other industries so as to develop cooperative ways of minimizing packaging and recycling materials.

These are lofty principles indeed, some unattainable at present. But the gauntlet has been thrown down.

The core tools of industrial ecology are life cycle analysis (LCA) and design for environment (DFE).

Life Cycle Analysis

The Society for Environmental Toxicology and Chemistry has defined LCA as follows:

> *[LCA] is an objective process to evaluate the environmental burdens associated with a product, process or activity by identifying and quantifying energy and material usage and environmental releases, to assess [their] impact and to evaluate and implement opportunities to effect environmental improvements. The assessment includes the entire life cycle of the product, process or activity, encompassing extracting and processing raw materials, manufacturing, transportation, and distribution; use/re-use/maintenance; recycling; and final disposal.*[5]

LCA deals primarily with existing products. The process is similar to two quality tools: the prioritization matrix and quality function deployment. Essentially LCA entails development of a series of interlocking matrices across three stages: inventory analysis, impact analysis, and improvement analysis. LCA is still embryonic, but international LCA standards are being developed. As LCA is refined it will become the core tool in the product and service control program.

Design for Environment

DFE deals with the improvement analysis stage of LCA and with products in the design stage. The process is the same as that for LCA: the development of interlocking matrices linking material inventories with impacts and possible improvements. It is beyond the scope of this book to describe LCA and DFE in detail. The interested reader is referred to Graedel and Allenby's excellent text.[6]

Significance of Industrial Ecology

Industrial ecology represents a major paradigm shift. It redefines the relationship between industry and the environment, and its implementation will be neither quick nor easy. It is safe to say that very few firms have begun to experiment with DFE, for example. Nonetheless, it represents a major breakthrough.

Industrial ecology is a direct bridge between environment and quality. It is an expression of the core quality concepts: systems thinking and upstream control. Its techniques have much in common with long-standing quality tools. Industrial ecology will inform environmental management for a long time and will make its synergy with quality ever more apparent.

The Cost of Safety and Environment

Estimating the cost of quality is a core activity in a quality system. Tracking the cost of safety and environment is equally important to the TSEM system. Safety and environment costs have generally been buried in overhead accounts. Neither industry nor society has known their magnitude.

As discussed in chapter 10, the accounting profession is presently developing corresponding accounting principles.[7] With time they will join the generally accepted accounting principles that guide the accounting profession. In the interim, other tools are available.

Full Cost Accounting

Full cost accounting provides a means of accounting for safety and environment costs.[8] FCA is not a precise science, nor is it widely practiced. The key constraints are

- The difficulty of finding accurate financial data
- An absence of guidelines analogous to GAAP

FCA of safety and environment entails identifying and quantifying four categories of cost associated with a process, project, or plant.

- Direct costs: capital, labor, and raw materials
- Hidden costs: inspections, investigations, monitoring, and legal support
- Contingent liability costs: remedial liabilities and fines
- Less tangible costs: labor relations, public relations, and goodwill

Direct Costs

Direct costs are

- Capital costs/depreciation, including
  - —Buildings
  - —Equipment purchase and installation
  - —Utility connections
  - —Project engineering
- Operating and maintenance expenses, including
  - —Materials
  - —Labor
  - —Utilities

Let us look at each of these costs in the context of safety and environment. The direct costs of an incident include

- Capital costs: replacing damaged buildings and equipment, reinstalling equipment, and clean-up costs

- Operating costs: lost production time and materials, labor costs including lost time of supervisors and coworkers aiding an injured worker or cleaning up a discharge, lost time associated with reorganizing and restarting the process

Direct costs can be quantified through traditional company accounting systems, depending on their sophistication.

Hidden Costs

Hidden costs comprise regulatory costs and costs that are unspecified or lumped into a general account. These include the cost of the following activities.

- Investigating incidents and correcting system weaknesses

- Selecting new employees

- Reporting, monitoring, attending hearings, and so on

- Legal representation

- Education and training

Contingent Liability Costs

Contingent liability costs are related to latent illnesses in workers or in the public caused by the company's operations. The size of this liability may be estimated from the following formula.

$$\text{Contingent liability} = \text{Probability of occurrence} \times \text{Estimated total cost}$$

Contingent liabilities are controversial because they can affect a company's balance sheet and, thus, its share value. In the context of safety and environment, estimating contingent liabilities requires an assessment of

- The degree of exposure to a given hazard
- The likely consequences to the health of workers or the public

Less Tangible Costs

These are the most difficult to quantify. They comprise intangibles such as the goodwill of the workforce and the community, good government, and community relations. These can lead to benefits such as

- Lower labor relations costs (fewer grievances and other disputes)
- Lower permitting and other government costs
- Lower legal and public relations costs

Or they can lead to the antithetical costs, of which perhaps the most important is market share lost due to negative consumer perceptions. This cost will be greater for companies that sell directly to the public than for suppliers. There are numerous examples of high-profile companies being boycotted because of poor environment and safety performance. The optics are never good in such cases.

Financial Measures

Decision makers typically use a number of traditional financial measures in determining the feasibility of an investment project. Such measures can be applied to the results of FCA analyses. These measures include

- Net present value
- Internal rate of return
- Profitability index
- Payback period

Since a discussion of these measures is beyond the scope of this book, the interested reader is referred to any basic text on financial management.

Summary

FCA is a difficult and imprecise process, but one worth doing, provided assumptions and expected degree of accuracy are noted. The absence of full information is no excuse for inaction. Whenever one is dealing with incomplete data, a range of assumptions should be made corresponding to a high, medium, and low continuum. Calculations made under each scenario will produce a range of results. This process allows one to dismiss some issues out-of-hand, prioritize other issues, and clarify one's thinking.

Characteristics of a TSEM Company

We are now in a position to describe a company practicing total safety and environmental management.

Senior management is actively involved in safety and environment. Senior managers participate in safety and environmental tours, inspections, investigations, and so on. Safety and environment are part of the agenda at senior management meetings. Managers with poor safety and environment records are rarely promoted, and many leave the company.

Safety and environment are invisible. Paradoxically, safety and environment are invisible because line managers accept them as part of their responsibilities. Absent are posters, contests, exhortations, and so on. Safety and environment are understood to be an integral part of quality. An incident is a serious nonconformity and evidence of a system weakness.

The safety and environment team's job in this environment is to

- Design and administer the system
- Track the cost of safety and environment
- Measure and assess the outputs of the system
- Audit the working of the system

Their underlying goal is to identify and eliminate variation.

Work conducted meets or exceeds best practice in the industry. Task analysis is used to reduce waste in work processes. Best safety and environment practices in the industry are used as benchmarks and adopted where feasible. Critical tasks and critical behaviors are clearly defined. The hiring and placement process ensures that workers have the knowledge, physical ability, and skills necessary to do their jobs. Environmental measurement is based on the best available benchmarks and the needs of the various audiences.

Decisions are based on data. Superstition, the bane of traditional safety management, gives way to decision making based on statistically valid measures of performance. Both upstream and downstream measures are employed. Managers are statistically literate.

Safety and environmental goals are SMART. Simple, measurable, achievable, reasonable, and trackable goals are set for the system as a whole and for its component programs. Managers are held accountable for safety and environmental performance. Safety and environment audits are conducted with the same rigor as financial audits. The PDCA cycle is used to achieve continual improvement.

The safety system's focus is on observable behaviors not worker attitudes. Management recognizes that the link between safety attitudes and safe behavior is tenuous. The system seeks to maximize the relaxed awareness of workers for optimal safety (and quality) performance. This is done by providing feedback in the form of highly visible charts of percent safe in the work area, and individual feedback provided by behavior observers.

The focus is on data. The emotion is taken out of safety. There are no exhortations to workers to work safely, no posters or contests.

Workers participate in safety and environmental efforts. Management recognizes that the knowledge and experience of the workforce is a vital resource to be utilized. Wage-roll employees know the traps, critical behaviors, idiosyncrasies of materials and equipment, and so on. They participate in incident investigations, inspections, and behavior observation.

Management tracks the cost of safety and the cost of environment. Full cost accounting keeps track of safety and environmental costs (aided by the sophisticated accounting system). Advances in environmental cost accounting are adopted as they appear.

Reported-but-not-incurred (RBNI) injuries are not an issue. Workers respond to the positive work environment. They like coming to work. The company's actions demonstrate that safety and environment are core company values on par with production. The resentment that is a root cause of the RBNI problem is absent.

Chapter Summary

We have described the TSEM system in detail, including both core and advanced programs of the SMS and EMS and their relationship to the policy. Broad program overlap has been found. Integration of these programs is eminently sensible.

The concept of industrial ecology was introduced in the context of the EMS. Industrial ecology redefines the relationship between industry and the environment. Its tools, LCA and DFE, are in their infancy but have links with quality tools. As LCA and DFE become refined, they will play an ever-increasing role in the EMS. Industrial ecology is a bridge between quality and environment. It expresses the core quality concepts of systems thinking and upstream control.

The accounting profession is developing environmental cost accounting principles. In the interim, the cost of environment (and safety) may be estimated using the tools of full cost accounting. Finally, the characteristics of a TSEM company are described.

Notes

1. J. M. Stewart, "The Multi-Ball Juggler," *Business Quarterly* (Western Business School, The University of Western Ontario) (winter 1993).

2. Jorgen Eklund, "Relationships Between Ergonomics and Quality in Assembly Work."

3. Takashi Osada, *The 5 S's: Five Keys to a Total Quality Environment* (Tokyo: Asian Productivity Organization, 1991).

4. T. E. Graedel and B. R. Allenby, *Industrial Ecology* (Englewood Cliffs, N.J.: Prentice Hall, 1995): 9.

5. Society for Environmental Toxicology and Chemistry, *A Technical Framework for Life-Cycle Assessment* (Washington, D.C.: SETAC, 1991).

6. Graedel and Allenby, *Industrial Ecology.*

7. The Canadian Institute of Chartered Accountants, *Reporting on Environmental Performance* (Toronto: CICA, 1994).

8. Global Environmental Management Initiative, *Finding Cost-Effective Pollution Prevention Initiative—A Primer* (Washington D.C.: GEMI, 1994).

CHAPTER 12

Dimensions of Synergy

To suggest is to create. To define is to destroy.

—Federico Garcia Lorca

It has not been my intention to rigidly define the synergy between quality, safety, and environment. I believe that they are self-evident and will develop on their own. Rather I have attempted to put a foundation beneath the intuitive understanding with which I began this book and to suggest the directions that these developments may take.

In this chapter we will discuss possible dimensions along which the synergy between quality, safety, and environment may develop. We will also summarize the key themes of the book.

Technological Synergy

Technological synergy refers to shared tools and techniques. I hope that this book will help to open a dialogue between these fields that will foster the transfer of technology.

The most obvious such synergy is the application of the quality approach to safety and environmental problems—the central theme of this book. Core quality concepts such as the following have herein been applied to safety and environmental management.

- The systems approach to management
- Statistical thinking and systems thinking
- Prevention or upstream control
- The quality triad (leadership, measurement, and participation)
- Deming's 14 points
- The quality toolbox

System safety tools such as fault tree analysis and failure mode and effect analysis (FMEA) are now part of the quality toolbox. They are equally applicable to environmental management. Safety tools such as behavior observation and task analysis also have great potential for quality and environmental management. In particular, defining and tracking the critical behaviors for quality may prove to be a rich field of endeavor.

Environmental tools are comparatively new and generally have not been widely applied outside the field. Tools such as life cycle assessment and design for environment are likely to have a growing impact on quality management as soon as they become standardized. And as professionals in these fields begin to talk to one another more technological synergy is likely to grow.

Structural Synergy—Systems

Structural synergy refers to the overlap between quality, safety, and environmental management systems. We have already discussed the broad overlap between the SMS and the EMS. There is also a degree of overlap between these and the quality management system (QMS).

Figure 12.1 displays a generic program menu for a QMS. Although we will not describe each program, some overlap is immediately obvious. Leadership is a common core program; task analysis, a common advanced program. The system maintenance and improvement

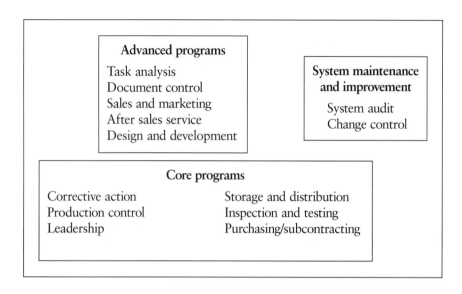

Figure 12.1. Quality management system generic program menu.

programs—system audit and change control—are identical with those in the SMS–EMS.

The most important overlapping program is leadership. When defining the quality policy it would make sense to refer to the safety and environment policies to ensure consistency. In addition, when defining the specific leadership responsibilities of managers it would be sensible to be aware of the corresponding safety and environment responsibilities.

Behavior observation and task analysis are powerful crossover programs that should be added to the quality tool box.

In chapter 11, I suggested that the EMS and SMS could be audited simultaneously—not a controversial idea. But the idea of integrating the system audit programs of the QMS and EMS–SMS has been contentious. International Organization for Standardization (ISO) Technical Committee 207, which was formed to help develop the ISO 14000 series of environmental standards, has been the scene of much impassioned debate over this issue. One side favored allowing integrated quality and environmental audits to save time and ease the burden on the organization. The other side believed that doing so would compromise the integrity of the environmental audits. The issue has yet to be resolved.

My thinking is that some form of audit integration is desirable, with the proviso that the auditors satisfy appropriate qualification standards. These would include auditing experience, as well as experience and professional credentials in both quality and environment/ safety. There would be comparatively few auditors that could meet these standards.

Finally, it should be noted that design and development and purchasing/subcontracting are likely to be strongly influenced by life cycle analysis and design for environment as these techniques mature.

In summary, there is a significant structural synergy between the quality, safety, and environmental management systems. This should be taken into account by system managers in the design and administration of these systems and during auditing so as to minimize the burden on the organization.

Political Synergy

> *You scratch my back and I'll scratch yours.*
>
> —Proverb

Importance of Political Behavior for Staff Groups

Power and *politics* are dirty words to some. Political behavior in organizations has traditionally been viewed as underhanded, opportunistic, or even immoral. Today, however, most management theorists take a more neutral view: Politics can be damaging if it diverts energy away from key organizational goals, but it can also play a vital role in achieving these goals.

Politics is the exercise of power by illegitimate means; that is, by circumventing the existing power structure. Institutionalized power tends to buffer the organization from reality. Political processes tend toward a realistic resolution of conflict among diverse interests. Thus, politics is a form of reality testing. It helps an organization stay in touch with its environment.[1]

How different the past 30 years might have been had General Motors, Ford, and Chrysler heeded the early warnings about off-shore competition! The message failed to get through the existing power structure. But politics might have had an effect by way of the backdoor.

Political behavior is essential for quality, safety, and environment professionals. They usually occupy staff positions in North American organizations and, thus, are vulnerable to staff versus line power imbalances. And as we have seen, the computer revolution tends to make commodities of professionals. They must recognize the obstacles they face and how learn to use politics to overcome them.

Political Strategies

Political strategies, therefore, are required to generate countervailing power to achieve important organizational goals.* The most effective strategies are alliance building and reciprocity. Alliance building is self-explanatory. Reciprocity means you scratch my back and I'll scratch yours.

It makes sense for quality, safety, and environment professionals to form alliances and practice reciprocity. Each group faces the same power imbalance with respect to the production bulls. Having allies who will support you at critical times (during the budgeting process, for example) will make it easier to achieve your goals.

---

*The interested reader is referred to Minztberg's classic, *Power In and Around Organizations* (Englewood Cliffs, N.J.: Prentice Hall, 1983).

Summary

Three dimensions are identified along which the synergy between quality, safety, and environment may develop. Technological synergy refers to shared tools and techniques. Structural synergy refers to system overlaps. Political synergy refers to the political strategies that professionals in these fields may adopt to overcome the endemic power imbalance each faces in most North American organizations.

Final Comments

> *Only when you have completed your journey will you understand what Ithaca means.*
>
> —Constantine Cavafy

This book has been a personal odyssey. I began it with a gut-level sense that there had to be a better way of managing safety and environment. As I complete it, I realize that I have been transformed. My absorption with the quality approach has changed the way I look at the world.

I have come full circle. Twenty years ago, *Zen and the Art of Motorcycle Maintenance* set off a personal quest for "the good." I believe I have found it, and as Cavafy understood, it was the journey that mattered.

Note

1. G. Salancik and J. Pfeffer, "Who Gets Power—And How They Hold On to It: A Strategic Contingency Model of Power," *Organizational Dynamics* (winter 1977).

Index